76-218

DEC 0 1 1976

D1596520

The person charging this material is responsible for its return on or before the **Latest Date** stamped below.

Theft, mutilation, and underlining of books are reasons for disciplinary action and may result in dismissal from the University.

D.C.S. Library

JUN 3 1977	FEB 1 3 1985
OCT 4 - 1977	May 24
NOV 3 1977	NOV 1 8 1988
JAN 1 9 1979	MAR 0 8 1994
	JAN 0 3 1995
MAR 2 1979	
MAR 2 8 1979	
JUN 1 1979	
AUG 0 9 1979	
OCT 1 2 1979	
MAR 6 1980	
OCT 2 5 1980	
FEB 3 1981	
OCT 2 6 1981	
JUL 2 0 1982	
AUG 6 1982	

(30088)

The Computational Complexity of Algebraic and Numeric Problems

ELSEVIER COMPUTER SCIENCE LIBRARY

Operating and Programming Systems Series
Peter J. Denning, Editor

1. *Halstead* A Laboratory Manual for Compiler and Operating System Implementation

Spirn Program Behavior: Models and Measurement (in prep.)

Programming Languages Series
Thomas E. Cheatham, Editor

1. *Heindel and Roberto* LANG-PAK— An Interactive Language Design System
2. *Wulf et al* The Design of an Optimizing Compiler
3. *Maurer* The Programmer's Introduction to SNOBOL

Cleaveland and Uzgalis Grammars for Programming Languages (in prep.)

Hecht Global Code Improvement (in prep.)

Theory of Computation Series
Patrick C. Fischer, Editor

1. *Borodin and Munro* The Computational Complexity of Algebraic and Numeric Problems

Computer Design and Architecture Series
Edward J. McCluskey, Editor

1. *Salisbury* Microprogrammable Computer Architectures
2. *Svobodova* Computer Performance Evaluation: Analysis and Application (in prep.)

Artificial Intelligence Series
Nils J. Nilsson, Editor

1. *Sussman* A Computer Model of Skill Acquisition

THEORY OF COMPUTATION SERIES   1

The Computational Complexity of Algebraic and Numeric Problems

A. Borodin
University of Toronto

I. Munro
University of Waterloo

American Elsevier Publishing Company, Inc.
New York   London   Amsterdam

AMERICAN ELSEVIER PUBLISHING COMPANY, INC
52 Vanderbilt Avenue, New York, N.Y. 10017

ELSEVIER PUBLISHING COMPANY
335 Jan Van Galenstraat, P.O. Box 211
Amsterdam, The Netherlands

ISBN 0-444-00156-5 (Paperbound Edition)
ISBN 0-444-00168-9 (Hardbound Edition)

©American Elsevier Publishing Co., Inc. 1975

All rights reserved.
No part of this publication may be reproduced,
stored in a retrieval system, or transmitted
in any form or by any means, electronic,
mechanical, photocopying, recording,
or otherwise, without permission in
writing from the publisher.
American Elsevier Publishing Company, Inc.
52 Vanderbilt Avenue, New York, N.Y. 10017.

Library of Congress Cataloging in Publication Data

Allan Borodin
  The Computational Complexity of Algebraic
  and Numeric Problems—A. Borodin, I. Munro

(Elsevier Computer Science Library: Theory of
Computation series; 1)
  Bibliography: p
  Includes index.
1. Computational Complexity I. Munro, Ian,
1947, joint author. II. Title.
QA76.B65
ISBN 0-444-00168-9
ISBN 0-444-00156-5 pbk.
DDC 519.7
LC 7421786

Manufactured in the United States of America

*To Judy, Jill, and Sara
whose complexities are unbounded.*

Contents

Preface ix

I. Introduction 1

II. Lower Bounds and Concepts Related to Linear Algebra 8

III. Lower Bounds and Concepts Related to Algebraic Independence 54

IV. The Fast Fourier Transform, Related Concepts, and Applications 77

V. Nonlinear Lower Bounds 107

VI. Parallel Processing in Numeric Computation 125

VII. The Complexity of Rational Iterations 148

Glossary of Notation 159

Bibliography and References 163

Index 173

PREFACE

This text is concerned with a field of research known as arithmetic or algebraic complexity. We view such research as a subfield of computational complexity or, more specifically, as a subfield of "concrete" computational complexity. We are thus interested in the inherent or intrinsic complexity of common arithmetic and algebraic computations, such as polynomial evaluation, interpolation, and matrix multiplication. To this end, we place our emphasis on establishing (that is, proving) lower bounds and on the construction of theoretically fast algorithms. We believe that the theoretical orientation of algebraic complexity—and indeed, of computational complexity—can be well justified, justified in terms of the conceptual insights made available for the more practical aspects of computing.

As the title suggests, the proof techniques are mainly algebraic; their level of mathematical formality varies throughout the text. We have tried to motivate informally the presentation, but we do rely on algebraic notation (mostly standard) to clarify possible ambiguities. A Glossary of Notation has been provided. It is expected that the reader is familiar with concepts from modern algebra. Some of the results discussed here may be of interest only to the "specialist." On a first reading, the thrust of the development should not be significantly impaired if one omits the proofs and details in Sections 2.2, 2.3, 2.4, 2.6, 3.3, 3.4, 6.2, and 7.2. The problems at the end of each chapter are used to provide examples as well as to complete some proofs. The open problems and conjectures serve to survey the present (that is, at the time of publication) state of the art.

During the past few years the area of "algebraic complexity" has been significantly influenced by a number of new results. We do not believe, nor hope, that the field has now "stabilized," but we do believe that an organization and assessment of the field will be beneficial at this time. In the Bibliography we have generally listed only the most recent and widely available reference for a given paper. This somewhat distorts the historical significance of certain papers (e.g., Winograd, 1970a). We hope that we have been fair in assessing credit for the various results.

Our text has evolved from notes developed in graduate courses given at the University of Toronto and the University of Waterloo. We were originally motivated by Richard Karp's Berkeley notes and have been considerably influenced by the students at Toronto and Waterloo. Specifically, we would like to acknowledge the helpful suggestions and corrections of many colleagues, including S. Cook, P. Fischer, N. Friedman, Z. Kedem, D. Kirkpatrick, H. Kung, A. Meyer, R. Moenck, L. Revah, V. Strassesn, and J. Traub.

Finally, we want to thank the National Research Council for its generous support throughout the past six years.

Toronto, Canada
January 1975

A. Borodin
I. Munro

Chapter 1

INTRODUCTION

Every programmer wants to write the best possible program for the problem being solved. "Best" may mean the program that uses the least storage, is easiest to understand, or reflect some other measure. Perhaps the most frequently used measure is *run time*, and that is the problem to which we will address ourselves. Given a function, what is the minimum number of machine cycles (on some reasonable computer) in which that function may be computed by any program?

1.1 Arithmetic Complexity—A Branch of Computational Complexity

Computational complexity is the study of "what makes functions hard to compute." Three rather different aspects have been undertaken in this general field of study. (For an overview of the entire field, see Borodin [1973a].) The first is an abstract approach to computing, which is furnished by a relatively machine-independent study of the theory of complexity. This work, based heavily on recursive function theory, provides basic results on the nature of computational complexity (on almost any computing device using almost any measure of complexity—time, space, length of program, etc.).

The second approach is to consider specific models of computation (e.g., Turing machines, random access machines, etc.), with respect to time and space bounds for computing various classes of functions. In general the results obtained say more about the power of the models of computation than they do about the difficulty of computing specific functions. There are important cases (e.g., the NP complete problems of Cook [1971] and Karp [1972]) in which the exact model of computation is not terribly important, and so the simplicity of a Turing machine is convenient for proving results. On the other hand, a Turing machine (Markov algorithm, etc.) seems a poor model of an "IBM 370" if we wish to study problems such as polynomial evaluation.

This leads us to the third area, the one of present interest, the study of efficient or optimal algorithms for the evaluation of specific "natural" functions on (idealized) digital computers. In this volume, we will restrict the "natural" functions under consideration to problems dealing with arithmetic computation and symbolic algebra. Nonnumeric computations, such as searching and sorting, pattern matching, and problems dealing with graph manipulation, although related, have a somewhat different flavor and will not be discussed here. (See Knuth [1973a, b] and Aho et al. [1974].)

Arithmetic complexity was, in fact, one of the first branches of computational complexity to be studied. Ostrowski (1954) and Motzkin (1955) proved some basic results on the number of arithmetics needed to compute polynomials. During the early 1960s, there were results concerning the complexity of iterative computations and some more results concerning polynomial evaluation. But it was not until Winograd's (1967) influential paper that "arithmetic or algebraic complexity" began to be recognized as a field of study. The single result that has thus far provided the greatest impetus to the field is an algorithm (more so than any lower bound proof) due to Strassen for multiplying matrices.

1.2 Intuition Can Be Wrong—Matrix Multiplication

Consider the problem of multiplying two n by n matrices. The standard algorithm requires n^3 multiplications and $n^3 - n^2$ additions. At first it may seem hopeless to attempt to reduce the number of arithmetics involved. Strassen (1969), however, observed that a pair of 2 by 2 matrices can be multiplied in 7 (not 8) multiplications by the following algorithm:

Intuition Can Be Wrong—Matrix Multiplication

ALGORITHM 1.2.1

Let
$$A = \begin{pmatrix} a_{11} & a_{12} \\ a_{21} & a_{22} \end{pmatrix}, \quad B = \begin{pmatrix} b_{11} & b_{12} \\ b_{21} & b_{22} \end{pmatrix}$$

and
$$AB = C = \begin{pmatrix} c_{11} & c_{12} \\ c_{21} & c_{22} \end{pmatrix}$$

Then we may find C by

$$P_1 = (a_{11} + a_{22})(b_{11} + b_{22})$$
$$P_2 = (a_{21} + a_{22}) b_{11}$$
$$P_3 = a_{11} (b_{12} - b_{22})$$
$$P_4 = a_{22} (-b_{11} + b_{21})$$
$$P_5 = (a_{11} + a_{12}) b_{22}$$
$$P_6 = (-a_{11} + a_{21})(b_{11} + b_{12})$$
$$P_7 = (a_{12} - a_{22})(b_{21} + b_{22})$$

Then

$$C_{11} = P_1 + P_4 - P_5 + P_7$$
$$C_{12} = P_3 + P_5$$
$$C_{21} = P_2 + P_4$$
$$C_{22} = P_1 + P_3 - P_2 + P_6$$

One's first reaction to this may be to say that trading 1 multiplication for 14 additions (this algorithm uses 18 additions; the usual one, 4) is not that great an idea in practice. The key point, however, is that the algorithm does not make use of the commutativity of multiplication. This means that the elements being multiplied can themselves be matrices, which gives rise to a recursive algorithm for multiplying matrices. Matrices of order 2^0 can be multiplied in 1 multiplication. Square matrices of order 2^{k+1} can be multiplied in 7 multiplications of order 2^k matrices (and 18 matrix additions). Let $M(n)$ denote the number of multiplications our algorithm uses to multiply two n by n matrices (if n is not a power of 2, pad the matrices out with zeros so that they are of this form). We may write

$$M(1) = 1, \qquad M(2^{k+1}) = 7M(2^k),$$

which implies $M(2^k) = 7^k$; setting $n = 2^k$,

$$M(n) = n^{\log_2 7} \qquad (\log_2 7 \cong 2.81).$$

(All logarithms henceforth will be to base 2 unless otherwise noted.)

If n is not a power of 2, the algorithm takes at most $7n^{2.81}$ multiplications; i.e., for very large n, substantially fewer than n^3 multiplications.

But what about the additions? An analysis of the algorithm shows that these, too, are bound by a constant times $n^{\log 7}$. Let $A(n)$ denote the number of additions in the algorithm. Then

$$A(1) = 0$$
$$A(2^{k+1}) = 18(2^k)^2 + 7A(2^k),$$

implying $A(2^k) = 6 \cdot 7^k - 6 \cdot 4^k \leqslant 6 \cdot 7^k$. And so,

$$A(n) \leqslant 7 \cdot 6 \cdot n^{\log 7}.$$

THEOREM 1.2.2

A pair of n by n matrices can be multiplied in $O(n^{\log 7})$ operations; i.e., in (at most) $c \cdot n^{\log 7}$ arithmetic operations for some constant c.

We have suggested using Algorithm 1.2.1 for multiplying 2 by 2 matrices. This requires 25 arithmetics, as opposed to 12 by the usual method. We could instead recurse down with the Strassen method only until we reach fairly small (say 8 by 8) matrices, and use the "standard algorithm" at this point. We also suggested embedding n by n matrices into ones of order $2^{\lceil \log n \rceil}$. If we just pad them up to $m2^k$ (for an appropriate choice of m and k) and "don't count" the multiplications and additions involving these "artificial 0's," the following corollary (also due to Strassen) follows.

COROLLARY 1.2.3

A pair of n by n matrices may be multiplied in $\leqslant 4.7n^{\log 7}$ arithmetic operations.

The Model of Computation

We are not suggesting that this algorithm be used in most practical instances (the overhead of the recursion and indexing would be prohibitive in most applications, not to mention round-off considerations). Strassen's method is, however, asymptotically faster than the usual one, and does make the point that matrix multiplication is not an n^3 process. It would seem likely that there are practical matrix multiplication algorithms using far fewer arithmetics. What is the minimum number of arithmetics that any algorithm can use to perform this process? Is it n^2, $n^2 \log n$, or indeed $n^{\log 7}$? How can we find such an algorithm? And what is even more interesting, how could we prove such an algorithm "optimal"?

The view we take here is an attempt to understand what makes functions hard to compute, rather than to provide the best practical algorithms. This view leads to a concern with asymptotic behavior and the presentation of some algorithms simply as "proofs of upper bounds," not as guidelines for practical computation.

1.3 The Model of Computation

If we are to demonstrate lower bounds or algorithms, for computing certain classes of functions, we must first state clearly what model of computation and what measures of complexity will be considered. Our underlying model of computation will be a random access register machine; the measure of difficulty, the number of instructions (of various types) executed. In evaluating arithmetic functions, it is easily seen that looping and branching instructions, although conceptually and practically useful, are often not necessary, as the code may be "unwound" and separate programs can be written for each "degree" of the desired class of functions. By this we mean, for example, that a program is written to evaluate polynomials of the specific degree k rather than giving k as a parameter for a more general program, as would be done in practice. In general, we will be concerned with the computation of rational functions; such functions can be realized by a sequence of instructions of the form $a \leftarrow b$ (op)c, where a is a variable, b and c are constants, inputs or previously computed variables, and (op) stands for $+$, $-$, $\times$, or $\div$. Arithmetics will be assumed to be exact, and so problems of round-off will be avoided.

In fact, we will formally think of our computations as computing elements of the ring $F[x_1, \ldots, x_n]$ or the field $F(x_1, \ldots, x_n)$, which denote respectively the polynomial ring and the rational field extensions of a field F by the indeterminates $\{x_1, \ldots, x_n\}$. For many of the results the choice of F will be arbitrary, but sometimes we will use properties of specific rings and fields such as **R** (the real field), **C** (the complex field), **Q** (the field of rationals), **Z** (the integers), or $\mathbf{Z}_p$ (the integers mod p). This area of study is often referred to as "algebraic complexity," reflecting the fact that the proof techniques are almost entirely algebraic.

We will adopt some of the terminology in Winograd (1970a) and say, "program P computes S in A given B." By this we mean that the program

$$P = \begin{matrix} s_1 \\ \cdot \\ \cdot \\ \cdot \\ s_t \end{matrix}$$

computes a set of elements $S = \{f_1, \ldots, f_r\}$ in the ring or field A, given the elements of B as inputs; i.e., there are $s_{i_1}, \ldots, s_{i_r}$ which (as elements of A) are identical to $f_1, \ldots, f_r$. For example, let $A = \mathbf{C}[x_1, x_2]$ and $B = \mathbf{C} \cup \{x_1, x_2\}$. Then

$$s_1 \leftarrow ix_2, \quad s_2 \leftarrow x_1 + s_1, \quad s_3 \leftarrow x_1 - s_1, \quad s_4 \leftarrow s_2 \times s_3$$

computes $s_4 = x_1^2 + x_2^2$. Step s_1 is a scalar multiplication, since $i \in \mathbf{C}$. We will see that any computation for $x_1^2 + x_2^2$ in $\mathbf{R}[x_1, x_2]$, given $\mathbf{R} \cup \{x_1, x_2\}$, requires at least 2 nonscalar multiplications (whereas the above computation over $\mathbf{C}[x_1, x_2]$ uses only 1 nonscalar multiplication).

One more example should help to clarify the symbolic nature of the model. Let F be any field. Now let $A = F(x)$ and $B = F \cup \{x\}$. We can compute $p(x) = \sum_{i=0}^{7} x^i$ in A, given B, by the program

$$s_1 \leftarrow x \times x, \quad s_4 \leftarrow 1 - s_3,$$
$$s_2 \leftarrow s_1 \times s_1, \quad s_5 \leftarrow 1 - x,$$
$$s_3 \leftarrow s_2 \times s_2, \quad s_6 \leftarrow s_4 \div s_5.$$

The Model of Computation

The fact that the program could not be executed for $x = 1$ is not an issue here; as elements of $F(x)$, $(1 - x^8) \div (1 - x)$ is just another representation for $1 + x + \cdots + x^7$. It is important to note that all lower bounds for $p(x) \in F[x]$ which we can derive using this model will apply to any program that is to correctly evaluate p on a sufficiently large set of points. Since we are mainly motivated by lower bounds, we are well justified in considering (symbolic) computations in $F(x_1, \ldots, x_n)$ even if we are only computing polynomials. If we specify that the computation is in $F[x_1, \ldots, x_n]$, we are implicitly saying that $\div$ is not allowed.

In Chapter 6 the model is modified somewhat as we focus on the evaluation of rational functions on a parallel processing machine. There are several reasonable extensions of the model previously described to parallel computation. We could model a vector processing machine or a pipeline machine. Instead, we will assume a very general model, that of k separate synchronous processors, each having access to the same storage. More precisely, at any parallel step i, any processor may use any input or any element computed by any processor before step i. By introducing all new variable names at each step, we insure that different processors will not attempt to store into the same register. We will refer to this model as *k-parallelism* (for k processors) and, for the case in which k is to be considered arbitrarily large, as *unbounded parallelism*.

We use the basic serial and parallel models to study the computation of rational functions. The evaluation of polynomial roots is not such a computation. In Chapter 7 we investigate this problem, and in doing so we move our mathematical basis toward analysis, although the proof techniques remain relatively algebraic. In that chapter we are concerned with the "efficiency" at which rational iterations over **R** or **C** are converging.

Chapter 2

LOWER BOUNDS AND CONCEPTS RELATED TO LINEAR ALGEBRA

The underlying theme of this chapter is that we need rely only on relatively simple ideas from linear algebra to successfully analyze some of the more basic problems confronting arithmetic complexity. The fact that the results are "to be expected" does not diminish the need for formal proofs. Throughout this chapter we will assume that F is any infinite field, although the reader may wish to think of F as the real or complex field.

2.1 Polynomial Evaluation, Substitution, and Active * Operations

The modern history of arithmetic complexity (assuming everything in these notes was not known to Gauss or Newton) begins with Ostrowski (1954). Ostrowski was the first to ask whether a good algorithm, Horner's rule for polynomial evaluation, was the "best method possible." Horner's rule, or synthetic division, is the polynomial evaluation algorithm suggested by writing $\Sigma\, a_i x^i$ as

$$(\ldots(a_n x + a_{n-1})x + a_{n-2})\ldots)x + a_0.$$

We see that this technique requires n multiplications and n additions. It avoids the $n-1$ multiplications needed for $\{x^i \mid i = 1, \ldots n\}$ in the more obvious algorithm. Ostrowski was only partially successful in answering the fundamental question he posed.

Polynomial Evaluation, Substitution, and Active * Operations

THEOREM 2.1.1

Every algorithm (using only +, −, ×) which computes $\sum_{i=0}^{n} a_i x^i$ in $F[x, a_0, \ldots, a_n]$, given $F[x] \cup \{a_i\}$, requires n ± operations. (The same result with ÷ allowed motivates Section 2.2.)

Proof. Any program P for $\sum_{i=0}^{n} a_i x^i$ can be transformed into a program P' for $\sum_{i=0}^{n} a_i$ in $F[a_0, \ldots, a_n]$, given $F \cup \{a_0, \ldots, a_n\}$. The new program P', which uses no more ± operations than P, is simply derived by preceding P with the statement $s \leftarrow 1$ and replacing every occurrence of x by s. For notational simplicity we can write $\Sigma a_i = \Sigma a_i x^i \mid_{x \leftarrow 1}$ and $P' = P \mid_{x \leftarrow 1}$. That is, the substitution $x \leftarrow 1$ changes both the program and the function being computed. We will prove by induction on n that $\sum_{i=0}^{n} a_i$ requires n ± operations. For $n = 1$, we wish to compute $a_0 + a_1$ as an element of $F[a_0, a_1]$. Without any ± operations, we can compute only elements of the form $c \prod_{i=0}^{l} a_i^{n_i}$, $c \in F$, $n_i \in \mathbb{N}$, the set of nonnegative integers. Thus, $a_0 + a_1$ requires one ± operation. For the induction step, let $\Pi_1(\vec{a}) \pm \Pi_2(\vec{a})$ be the first "essential" addition, where $\Pi(\vec{a})$ represents an element of the form $c \prod_{i=0}^{n} a_i^{n_i}$. Without loss of generality, assume that a_n appears in Π_1 and make the substitution $a_n \leftarrow 0$. In particular, we have eliminated the first ± operation. More precisely, we can eliminate this operation in the new program, since its result is the constant 0. We are now computing $\sum_{i=0}^{n-1} a_i$, which by the induction hypotheses requires $(n-1)$ ± operations. □

While this proof is straightforward, it does serve to motivate the idea of a *substitution argument*. From Theorem 2.1.1 it is seen that Horner's rule is an optimal way of evaluating a polynomial with respect to the number of additions used. We will now show (by a somewhat more involved substitution argument) that it is also optimal with respect to the number of multiplications/divisions used, and hence optimal with repsect to total arithmetics.

Definition 2.1.2

Suppose $F \subseteq G$ and that we are computing in $G(a_1, \ldots, a_n)$. The operation $f * g$ (where $*$ denotes $\times$ or $\div$) is said to be *inactive relative to F and G* (we will just say *inactive* when the context is clear) if one of the following holds:
(i) $g \in F$; i.e., g is a "constant."
(ii) $f \in F$ and the operation is $\times$.
(iii) $f \in G$ and $g \in G$.

Throughout the remainder of this section, we will be interested in computations in $G(a_1, \ldots, a_n)$, given $G \cup \{a_i\}$, where $G = F(x)$. Here the inactive operations are either *scalar multiplications* or just involve rational functions of x. Any $\times$ or $\div$ which is not inactive is said to be *active*. Thus, $x \times x$, $4 \times a_1$ are inactive, while $x \times a_1$, $a_1 \times a_2$, and $4 \div a_1$ are active. The following lemma is central to the basic active substitution argument.

Lemma 2.1.3

Let $l(a_2, \ldots, a_s)$ denote a linear combination $\Sigma\, c_i a_i$ with $c_i \in F$. Suppose $l_1(a_1, \ldots, a_s), \ldots, l_u(a_1, \ldots, a_s)$ are linearly independent over F. Now let $a_1 = l(a_2, \ldots, a_s)$ and $l'_i(a_2, \ldots, a_s) = l_i|_{a_1 \leftarrow l(\vec{a})} = l_i(l(a_2, \ldots, a_s), a_2, \ldots, a_s)$. Then there are at least $u - 1$ linearly independent l'_i.

Proof. Think of the $\{a_i\}$ as generating an s dimensional vector space S over F, and consider $l_1, \ldots, l_u$ to be u linearly independent vectors in S. Let S_1 be the u dimensional subspace generated by the $\{l_i\}$. Now consider the linear mapping ϕ induced by $\phi(a_1) = l(a_2, \ldots, a_s)$ and $\phi(a_i) = a_i$, $2 \leq i \leq s$. The kernel of ϕ has dimension 1, and therefore the kernel of ϕ restricted to S_1 has dimension at most 1. Then $\phi(S_1)$ is the space generated by $\{l'_1, \ldots, l'_u\}$ and has dimension equal to $[\dim S_1 - \dim \ker \phi|_{S_1}] \geq u - 1$. Of course the substitution can be for any a_j rather than for just a_1. $\square$

Theorem 2.1.4

Any computation in $F(\vec{a}, x)$, given $F(x) \cup \{a_i\}$ of an expression of the form

$$\sum_{i=0}^{n} l_i(\vec{a})x^i + r(x) + l_0(\vec{a}),$$

in which there are u linearly independent terms in $\{l_i(\vec{a}) \mid 1 \leq i \leq n\}$ and $r(x)$ is a rational function of x, requires at least u active operations.

Proof (by induction on u). Without an active operation, we can compute only elements of the form $l(\vec{a}) + r(x)$ with $r(x) \in F(x)$. Hence, the theorem is true for $u = 1$. Assume the theorem is true for $u = 1, \ldots, m$ but is false for $u = m + 1$. The proof technique is similar to that of Theorem 2.1.1; i.e., substitute for some a_i and see what the algorithm then computes. Suppose the contradictory scheme uses only $k < m + 1$ active operations, and let $[\hat{l}_1(\vec{a}) + r_1(x)] * [\hat{l}_2(\vec{a}) + r_2(x)]$ be the first active operation. Since the operation is active, we know that either $\hat{l}_1 \notin F$ or $\hat{l}_2 \notin F$, and hence we can assume without loss of generality that $\hat{l}_2(\vec{a}) = \sum_{i=1}^{s} c_i a_i$ with (say) $c_1 \neq 0$. We want to make a substitution for a_1 after which $\hat{l}_2(\vec{a}) + r_2(x)$ will be a member of F. It suffices to make the substitution $a_1 \leftarrow \tilde{l}(\vec{a}) + \tilde{r}(x) = -1/c_1 (\sum_{i=2}^{s} c_i a_i + r_2(x)) + c$ for any $c \in F$. (The constant c is chosen so that the substitution will not result in any division having 0 as the divisor. By assuming that F is infinite, it is always possible to find such a $c \in F$.) The substitution $a_1 \leftarrow \tilde{l}(\vec{a}) + \tilde{r}(x)$ defines a new program, which computes

$$\sum_{i=1}^{n} l_i(\vec{a})x^i + l_0(\vec{a}) + r(x) \Big|_{a_1 \leftarrow \tilde{l}(\vec{a}) + \tilde{r}(x)}.$$

This expression can be rewritten in the form

$$\sum_{i=1}^{n} l'_i(\vec{a})x^i + l'_0(\vec{a}) + r'(x).$$

In the new program, a_1 will be replaced by the expression $s_1 = \tilde{l}(\vec{a}) + \tilde{r}(x)$, which is computed at no cost in active operations. What was previously the first active * operation is no longer an active operation, and any operation that was previously inactive remains inactive. Hence, the new program uses at most $k-1 < m$ active operations to compute $\sum_{i=1}^{n} l'_1(\vec{a})x^i + l'_0(\vec{a}) + r'(x)$. But $\{l'_i(\vec{a})\} = \{l_i(\vec{a})|_{a_1 \leftarrow \bar{i}(\vec{a})}\}$, and thus there are at least m linearly independent terms in $\{l'_i \mid 1 \leq i \leq n\}$, contradicting the induction hypothesis.

$\square$

COROLLARY 2.1.5 (Pan 1966.]

Horner's rule is an optimal algorithm with respect to the number of multiplications/divisions for the evaluation of $\sum_{i=0}^{n} a_i x^i$.

Proof. It follows from the theorem (since $\{l_i(\vec{a}) = a_i \mid 1 \leq i \leq n\}$ are linearly independent) that n active operations are required. Indeed, as is often the case, it is the corollary that suggests the theorem. For, in trying to apply the substitution $a_1 \leftarrow \bar{l}(\vec{a}) + \bar{r}(x)$ to $\Sigma a_i x^i$, we were "forced" into the more general form $\Sigma l_i(\vec{a})x^i + r(x)$ in order to apply an induction argument.

$\square$

2.2 Basic Substitution Argument for Additions/Subtractions

It remains to remove the restriction (no $\div$) from the $\pm$ lower bound of Theorem 2.1.1. For polynomial evaluation, this turns out to be a corollary of the $\pm$ lower bound with preconditioning (see Section 3.2). But for completeness, it would seem that we should be able to prove $\sum_{i=0}^{n} a_i$ requires $n \pm$ operations while allowing $\times$ and $\div$. The difficulty with a simple substitution argument is that the first $+$ operation might be something like $(a_1 \div a_2) + (a_2 \div a_1)$, which would preclude setting either a_1 or a_2 to zero.

We will now sketch some ideas due to Kedem and Kirkpatrick (1974). This development will yield the correct bound for $\sum_{i=0}^{n} a_i$ and will also provide lower bounds on $\pm$ operations for a number of other problems. The crux of this development is that we need an appropriate notion of "independence" of terms. Having established such a notion, the proof will proceed by induction on the number of independent terms. The inductive step will depend (as before) on a substitution argument. And just as in Theorem 2.1.4, the substitution "forces" a generalization of the specific problem (in this case Σa_i).

Basic Substitution Argument for Additions/Subtractions

DEFINITION 2.2.1

Let $Q\langle a_1, \ldots, a_s\rangle$ denote the set of all *generalized monomials*, terms of the form $\prod_{i=1}^{s} a_i^{c_i}$ with $c_i \in Q$. $F[Q\langle a_1, \ldots, a_s\rangle]$ and $F(Q\langle a_1, \ldots, a_s\rangle)$ will then denote the ring and field extensions of F by $Q\langle a_1, \ldots, a_s\rangle$; i.e.,

$$F[Q\langle a_1, \ldots, a_s\rangle] = \left\{ \sum_{i=1}^{n} d_j \prod_{i=1}^{s} a_i^{c_{ij}} \,\Big|\, d_j \in F, c_{ij} \in Q \right\}.$$

Here $Q\langle a_1, \ldots, a_s\rangle$ may be viewed as a vector space over Q via the 1-1 correspondence $m = \prod_{i=1}^{s} a_i^{c_i} \leftrightarrow \langle c_1, \ldots, c_s\rangle$. Note that multiplication in $Q\langle \vec{a}\rangle$ corresponds to vector addition in the vector space and that raising an element to a rational power corresponds to scalar multiplication in the vector space. Abusing notation, we shall let m denote either a monomial or the corresponding vector, depending on the context.

A vector space can be defined in terms of a basis (a maximal set of independent vectors), with every vector in the space being representable as some linear combination of the basis vectors. A vector subspace can be defined as the space generated by some subset of vectors in the space. Geometrically, a vector subspace is a linear subspace passing through the origin. The (affine) dimension of a linear subspace that does not necessarily include the origin is defined as the (vector space) dimension of the space when translated to pass through the origin.

We may also speak of the affine dimension of any set of points W as being that of the smallest linear subspace including W. For example, a line or any two points have affine dimension equal to 1, whereas the vector space dimension or rank of any two independent vectors (points not lying on a line through the origin) is 2.

DEFINITION 2.2.2

Let $p = \sum_{i=1}^{t} c_i m_i$ be in $F[Q\langle a_1, \ldots, a_s\rangle]$, and let y be a new indeterminate so that py is in $F[Q\langle a_1, \ldots, a_s, y\rangle]$. We let $\bar{p}$ denote $\{m_1, \ldots, m_t\}$ and define the "affine independence" of p as

$$\text{adim}(\bar{p}) \stackrel{\circ}{=} \text{vector space dimension}(\overline{py}) - 1$$
$$\stackrel{\circ}{=} \text{dimension}(\{m_1 y, \ldots, m_t y\}) - 1.$$

If S is a vector space over F and $W \subseteq S$, then $W^* \triangleq \{w \mid w = \Sigma \lambda_i w_i, w_i \in W, \Sigma \lambda_i = 1\}$ is the linear subspace generated by W. Multiplying by the indeterminate y (i.e., adding in the vector space) insures that $\overline{py}$ generates a linear space that does not pass through the origin so that $\operatorname{adim}(\overline{py})$, or dimension $(\overline{py}) - 1$, is the affine dimension of $(\overline{py})^*$. We now have an analogue to Lemma 2.1.3.

LEMMA 2.2.3

If $\operatorname{adim}\{m_1, \ldots, m_r\} = u$ and $m'_i = m_i\big|_{a_1 \leftarrow \hat{m}}$ for any $\hat{m} \in S$, then $\operatorname{adim}\{m'_1, \ldots, m'_r\} \geq u - 1$.

Proof. The proof is just a slight extension of Lemma 2.1.3. In that lemma we used the fact that (vector space) dimension $(\phi l_1, \ldots, \phi l_r) \geq$ dimension $(l_1, \ldots, l_r) - \ker \phi$ for any linear transformation ϕ. The same result holds for affine dimension, for we have

$$\begin{aligned}
\operatorname{adim}\{\phi m_1 y, \ldots, \phi m_r\} &= \text{dimension } \{\phi m_1 y, \ldots, \phi m_r y\} - 1 \\
&= \text{dimension } \phi(\{m_1 \phi^{-1} y, \ldots, m_r \phi^{-1} y\}) - 1 \\
&= \text{dimension } \phi(\{m_1 \hat{y}, \ldots, m_r \hat{y}\}) - 1, \\
&\quad \text{where } \hat{y} = \phi^{-1}(y) \text{ can be considered to be} \\
&\quad \text{another indeterminate} \\
&= \text{dimension } \{m_1 \hat{y}, \ldots, m_r \hat{y}\} - \ker \phi - 1 \\
&= \operatorname{adim}\{m_1, \ldots, m_r\} - \ker \phi
\end{aligned}$$

□

THEOREM 2.2.4

Let

$$p = \sum_{j=1}^{r} d_j m_j \in F[Q\langle a_1, \ldots, a_s \rangle],$$

where $d_j \in F$, m_j is a generalized monomial in $Q\langle \vec{a} \rangle$, and let $u = \operatorname{adim}(m_1, \ldots, m_j)$. Then any algorithm that computes p in $F^-(Q\langle a_1, \ldots, a_s \rangle)$, given $F \cup Q\langle a_1, \ldots, a_s \rangle$ requires at least $u \pm$ operations.

Proof. The proof will proceed by induction on u. Without a $\pm$ operation, we can only compute something of the form dm with $d \in F$ and $m \in Q\langle a_1, \ldots, a_s \rangle$; but $\operatorname{adim}(m) = 0$. This establishes the theorem for $u = 1$. Assume the theorem is true for $u = k$ and that $\operatorname{adim}(m_1, \ldots, m_r) = k + 1$. Suppose a program P computes p, using at most k

Generalizations of the Basic ~~Subtraction~~ Substitution Arguments

± operations, and that $s_1 \pm s_2$ is the first such operation, with $s_i = d_i m_i$. We can modify the program in the following way: $s_1 \pm s_2$ is replaced by: s_1/s_2; $s_1/s_2 \pm 1$; $s_2(s_1/s_2 \pm 1)$. So without loss of generality we can assume that the first ± operation is $(d \prod_{i=1}^{s} a_i^{c_i}) \pm 1$. We want to make a substitution so that $d \prod_{i=1}^{s} a_i^{c_i} = d' \in F$, thereby eliminating the need for the ± operation. Assuming $c_1 \neq 0$, it suffices to set $a_1 \leftarrow (d'/d) \prod_{i \neq 1} a_i^{-c_i/c_1} \triangleq \hat{dm}$. Then, using less than k ± operations, we can compute $p|_{a_1 \leftarrow \hat{dm}}$. But $p|_{a_1 \leftarrow \hat{dm}} = \sum_{i=1}^{r} d'_i m'_i$, where $m'_i = m_i|_{a_1 \leftarrow \hat{m}}$. By Lemma 2.2.3, $\{m'_1, \ldots, m'_r\}$ has adim $\geqslant k$. □

COROLLARY 2.2.5

The computation of $\sum_{i=0}^{n} a_i$ in $F(a_0, \ldots, a_n)$, given $F \cup \{a_i\}$, requires n ± operations.

COROLLARY 2.2.6

The computation of $\sum_{i=0}^{n} a_i b_i$ in $F(\vec{a}, \vec{b})$, given $F \cup \{a_i\} \cup \{b_i\}$ requires n ± operations.

2.3 Generalizations of the Basic Substitution Arguments

The substitution arguments of Sections 2.1 and 2.2 can be applied to a number of different problems. Winograd (1970a) has developed an elegant framework to establish lower bounds for many arithmetic problems. We motivate this development by looking at polynomial evaluation again. By considering the number of linearly independent terms $\{l_i\}$, we derived a lower bound for $\sum_{i=0}^{n} l_i(\vec{a}) x^i + r(x)$. This could also be expressed as $\sum_{i=0}^{n} a_i r_i(x) + r(x)$; i.e., the sum of a vector inner product and a single term:

$$(r_0(x)\, r_1(x) \cdots r_n(x)) \begin{pmatrix} a_0 \\ \vdots \\ a_n \end{pmatrix} + (r(x)) = \Phi_{1 \times (n+1)} \vec{a} + \phi.$$

This generalizes nicely to a number of problems that involve computing t expressions $\{\psi_1 \mid 1 \leq i \leq t\}$ of the form

$$\begin{pmatrix} \psi_1 \\ \vdots \\ \psi_t \end{pmatrix} = \Phi_{t \times n} \begin{pmatrix} a_1 \\ \vdots \\ a_n \end{pmatrix} + \vec{\phi}_{t \times 1},$$

the sum of a matrix vector product and a vector. The elements of Φ and ϕ are in some field G (e.g., $G = F(x)$) and the $\{a_i\}$ are indeterminates. We then extend an appropriate version of linear independence to the columns of Φ and derive some general results.

DEFINITION 2.3.1

We say $e_1, \ldots, e_r$ are *linearly independent over F mod H* if there is no nontrivial $\{c_i\}$ in F such that $\sum_{i=1}^{r} c_i e_i$ is in H. Let the elements of $\Phi_{t \times n}$ be in G and let F be a subset of G; the columns $\Phi_1, \ldots, \Phi_u$ will be called linearly independent *with respect to F* if $\Phi_1, \ldots, \Phi_u$ are linearly independent over F mod $F_{t \times 1}$.

THEOREM 2.3.2 (Winograd)

Suppose we want to compute $\vec{\psi}_{t \times 1} = \Phi_{t \times n} \vec{a}_{n \times 1} + \vec{\phi}_{t \times 1}$ in $G(a_1, \ldots, a_n)$, given $G \cup \{a_1, \ldots, a_n\}$, where the elements of Φ and ϕ are in G. If there are u linearly independent columns $\{\Phi_i\}$ with respect to some infinite subfield $F \subseteq G$, then at least u active (relative to F and G) $\times, \div$ operations are required to compute the $\{\psi_i\}$.

Proof. The proof is similar to that of Lemma 2.1.4; namely, an induction on u with the induction step based on a substitution $a_i = l(\vec{a}) + g$ with g in G. After such a substitution (say, $a_n = \sum_{i=1}^{n-1} c_i a_i + g$), we are computing

$$\vec{\psi}'_{t \times 1} = \Phi'_{t \times (n-1)} \begin{pmatrix} a_1 \\ \vdots \\ a_{n-1} \end{pmatrix} + \vec{\phi}'_{t \times 1},$$

where $\Phi'_j = \Phi_j + c_j \Phi_n$. The reader can now verify that there are at least $u - 1$ linearly independent (with respect to F) columns $\{\Phi'_j\}$ □

COROLLARY 2.3.3

(a) Horner's rule minimizes the number of active $*$ operations for computing

$$\sum_{i=0}^{n} a_i x^i = (1 \quad x \cdots x^n) \begin{pmatrix} a_0 \\ \vdots \\ a_n \end{pmatrix} + 0.$$

(b) Evaluation of several polynomials at a single point is best computed by repeated application of Horner's rule. That is, $\sum_{i=0}^{n_j} a_{ij} x^i$ ($1 \leq j \leq t$), or equivalently

$$\begin{pmatrix} 1 \quad x \cdots x^{n_1} & & \\ & 1 \quad x \cdots x^{n_2} & \\ & & 1 \quad x \cdots x^{n_t} \end{pmatrix} \begin{pmatrix} a_{01} \\ \vdots \\ a_{n_1 1} \\ \vdots \\ a_{n_t t} \end{pmatrix} + \begin{pmatrix} 0 \\ \vdots \\ 0 \end{pmatrix},$$

requires $\sum_{j=1}^{t} n_j$ active $\times, \div$.

(c) The usual matrix vector computation is optimal, since $A_{r \times s} \vec{b}_{s \times 1}$, or equivalently

$$\begin{pmatrix} b_1 \cdots b_s & & \\ & b_1 \cdots b_s & \\ & & b_1 \cdots b_s \end{pmatrix} \begin{pmatrix} a_{11} \\ a_{12} \\ \vdots \\ a_{21} \\ \vdots \\ a_{rs} \end{pmatrix}$$

requires rs active $\times, \div$.

Although this framework is quite elegant, it must be remembered that the proof technique is based on a substitution for the $\{a_i\}$ and hence the derived lower bound cannot exceed the number of elements in $\{a_i\}$. Thus, to substantially improve the bound for problems like $A_{n \times n} B_{n \times n}$, we must expect to find a different proof technique.

Another generalization of the active operation-substitution technique is due to Strassen (1972). Strassen's framework and results are still restricted to the number of indeterminates a_i. The benefit of his framework is that it enables one to discuss the computation of functions in the field extension $G(\vec{a})$ rather than just in the linear extension of G with respect to $\vec{a}$. As before, let $F \subseteq G$.

THEOREM 2.3.4 (Strassen)

Suppose we can compute $\{f_1, \ldots, f_r\}$ in $G(a_1, \ldots, a_n)$, using $m \leq n$ active $\times, \div$ operations. (It helps to consider $r = 1$ and $m = n - 1$.) Then there is a dense subset D (i.e., a set not contained in the zeros of a nontrivial multivariate polynomial) of G^n with the following property: For every $\vec{v}$ in D there are matrices $\Gamma_{n \times (n-m)}$ over F, $\Lambda_{r \times (n-m)}$ over F, and $\vec{w}_{r \times 1}$ over G, such that Γ has rank

$$\begin{pmatrix} f_1(\Gamma\vec{t} + \vec{v}) \\ \vdots \\ f_r(\Gamma\vec{t} + \vec{v}) \end{pmatrix} = \Lambda\vec{t} + \vec{w} \qquad \vec{t} = \begin{pmatrix} t_1 \\ \vdots \\ t_{n-m} \end{pmatrix}$$

$(n - m)$, and the t_i are indeterminate with respect to G. That is, on a particular $n - m$ dimensional linear subspace through $\vec{v}$, the set o of functions are linear affine functions over F (i.e., $\Sigma c_i a_i + g$ with c_i in F, g in G).

Outline of Proof. We will just try to sketch the geometrical motivation. Suppose we are computing a function that needs only one multiplication: $[\sum_{i}^{n} c_i a_i + g_1] \times [\Sigma d_i a_i + g_2]$ with the g_i in G and the c_i, d_i in F. Then on certain $n - 1$ dimensional subspaces (i.e., defined by a hyperplane $a_i = l(\vec{a}) + g$), the function is linear affine with respect to the $\{a_i\}$. Similarly, for 2 multiplications, there are $n - 2$ dimensional linear subspaces, etc. Any linear subspace of dimension k is the range of some linear affine map: $G^k \to G^n$ of rank k.

□

Generalization of the Basic Substitution Arguments

COROLLARY 2.3.5

$f(a_1, \ldots, a_n) = \sum_{i=1}^{n} a_i x^i$ requires n active $*$ operations.

Proof. If the number of active $\times$, $\div$ is $m = n - 1$, then there exist $\{v_i\}$ in G, and $\{\gamma_i\}$, λ in F such that

$$\sum_{i=1}^{n} (\gamma_i t + v_i) x^i = (\sum_{i=1}^{n} \gamma_i x^i) t + \sum_{i=1}^{n} v_i x^i = \lambda t + w,$$

where x is to be considered as an indeterminant over F, $G = F(x)$, and t an indeterminant over G. Then $\Sigma \gamma_i x^i = \lambda$, ($\lambda \in F$), implies $\gamma_i = 0$ for all i, and thus

$$\Gamma = \begin{pmatrix} \gamma_1 \\ \vdots \\ \gamma_n \end{pmatrix}$$

does not have the required rank $(n - m = 1)$.

□

Strassen also proves the following corollary. (Now let $F = G$.)

COROLLARY 2.3.6

Let

$$f = \sum_{1 \leq i, j \leq n} \beta_{ij} a_i a_j$$

$$= (a_1, \ldots, a_n)\, (\beta_{ij}) \begin{pmatrix} a_1 \\ \vdots \\ a_n \end{pmatrix}$$

$$= \vec{a} B \vec{a}^T$$

If the dimension of a maximal null space of B is p, then $n - p$ active $\times$, $\div$ operations are necessary and sufficient to compute f. In particular, if $F = \mathbf{R}$ and $f = \sum_{i=1}^{n} a_i^2$, then $n \times$, $\div$ operations are required.

Sketch of Proof. It is known that

$$f \sim a_1 a_2 + \cdots + a_{2p-1} a_{2p} + \alpha_{2p+1} a_{2p+1}^2 + \cdots + \alpha_n a_n^2;$$

the similarity relation ($\sim$) is defined by $f = \vec{a}B\vec{a}^T \sim \vec{a}CBC^T\vec{a}^T = (\vec{a}C)B(\vec{a}C)^T$ for some nonsingular scalar matrix C. Since we do not count multiplication by constants in F, the transformation aC does not change the complexity. Hence $n - p$ operations are sufficient. Let m be the number of required $\times, \div$. By applying the theorem and equating coefficients, there exists $\{\gamma_{kl} \mid 1 \leq k \leq n, 1 \leq l \leq n - m\}$: for all l, $\sum_i \sum_j \beta_{ij} \gamma_{il} \gamma_{jl} = 0$. Hence, $\{(\gamma_{1l}, \ldots, \gamma_{nl})^T \mid 1 \leq l \leq n - m\}$ generates a null space for B and so $p \geq n - m$.

□

(Vari [1972] has independently obtained optimal results for quadratic forms over the real and complex fields.)

We conclude this section by giving Kedem and Kirkpatrick's (1974) extension of their basic ± argument to several rational functions. In doing so, some lemmas of independent interest are established. (Actually, the development given here subsumes the discussion in Section 2.2; that presentation was given as a parallel to Section 2.1). We first note that there is a relatively easy way to discuss the ± complexity of a set of functions.

FACT 2.3.7

If $\{f_1, \ldots, f_r\}$ is computable within t ± operations, then $f_1 + \cdots + f_r$ (or $f_1 y_1 + \cdots + f_r y_r$, where the y_i are new indeterminates) is computable within $(t + r - 1)$ ± operations. Thus, if $f_1 + \cdots + f_r$ requires k ± operations, $\{f_i \mid 1 \leq i \leq r\}$ requires $(k - r + 1)$ ± operations.

COROLLARY 2.3.8 (Kirkpatrick)

$A_{p \times n} B_{n \times q} = C_{p \times q} = (c_{ij})$ requires at least $(p + q - 1)(n - 1)$ ± operations.

Proof.

$$\text{adim}\left(\sum_{i=1}^{p} c_{i1} + \sum_{j=2}^{q} c_{1j}\right) = (p + q - 1)n - 1.$$

Recalling the notation from Section 2.2 (in particular, the association between monomials and vectors), Theorem 2.2.4 says that the computation of p requires at least adim($\bar{p}$) ± operations. We want to

Generalization of the Basic Substitution Arguments

extend the definition of affine independence to several functions so that the computation of $\{p_1, \ldots, p_r\}$ requires "a-indep $\{p_1, \ldots, p_r\}$" $\pm$ operations.

DEFINITION 2.3.9 (see Definition 2.2.1)

Let p_i be in $F[Q\langle a_1, \ldots, a_r \rangle]$, $1 \leq i \leq r$, and let $y_1, \ldots, y_r$ be new indeterminates so that $\Sigma p_i y_i$ is in $F[Q\langle a_1, \ldots, a_s, y_1, \ldots, y_r \rangle]$.
Define a-indep $(\{p_1, \ldots, p_r\}) \overset{\circ}{=}$ vector space of dimension $(p_1 y_1 + \cdots + p_r y_r) - r$.

In order to establish the desired result, we need two basic properties describing how a-indep is affected by the operations $+, -, \times$.

LEMMA 2.3.10

(a) a-indep $(\{p_1, p_2, \ldots, p_r, p_1 \times p_2\}) = a$-indep $(\{p_1, p_2, \ldots, p_r\})$.

(b) a-indep $(\{p_1, p_2, \ldots, p_r, p_1 \pm p_2\}) \leq a$-indep $(\{p_1, p_2, \ldots, p_r\}) + 1$.

Proof

(a) Let m_1° in p_1 and m_2° in p_2 be fixed and let $m_1 m_2 y_{r+1}$ be an arbitrary element of $(p_1 \times p_2) y_{r+1}$. Then $m_1 m_2 y_{r+1}$ can be expressed as $(m_1^\circ m_2^\circ y_{r+1})(m_1 y_1)(m_1^\circ y_1)^{-1}(m_2 y_2)(m_2^\circ y_2)^{-1}$, so that $m_1 m_2 y_{r+1}$ is in the vector space generated by $p_1 y_1 \cup p_2 y_2 \cup \{m_1^\circ m_2^\circ y_{r+1}\}$. Therefore,

a-indep $(\{p_1, \ldots, p_r, p_1 \times p_2\})$
$= $ dimension $(p_1 y_1 + p_2 y_2 + \cdots p_r y_r \cup (p_1 \times p_2) y_{r+1}) - (r+1)$
$= $ dimension $(p_1 y_1 + \cdots + p_r y_r \cup m_1^\circ m_2^\circ y_{r+1}) - (r+1)$
$= $ dimension $(p_1 y_1 + \cdots + p_r y_r) - r$ (since y_{r+1} is a new indeterminate, $m_1^\circ m_2^\circ y_{r+1}$ must be linearly independent with respect to $p_1 y_1 + \cdots + p_r y_r$)
$= a$-indep $(\{p_1, \ldots, p_r\})$.

(b) Observe that an arbitrary element $m_1 y_{r+1}$ in $p_1 y_{r+1}$ can be expressed as $(m_1^\circ y_{r+1})(m_1 y_1)(m_1^\circ y_1)^{-1}$, and $m_2 y_{r+1}$ in $p_2 y_{r+1}$

can be expressed as $(m_2^o y_{r+1})(m_2 y_1)(m_2^o y_1)^{-1}$. Therefore, $(p_1 + p_2)y_{r+1}$ can be generated by elements of $p_1 y_1 \cup p_2 y_2 \cup \{m_1^o y_{r+1}\} \cup \{m_2^o y_{r+1}\}$. The proof now follows as in part (a). □

THEOREM 2.3.11

Let $\{p_1, \ldots, p_r\} \subseteq F[Q\langle a_1, \ldots, a_r \rangle]$ with a-indep $\{p_1, \ldots, p_r\}$ = u. Then at least $u \pm$ operations are required to compute $\{p_1, \ldots, p_r\}$ in $F[Q\langle a_1, \ldots, a_r \rangle]$, given $F \cup Q\langle a, \ldots, a_r \rangle$.

Proof. By induction on the total number of $+, -, \times$ operations, using Lemma 2.3.10 for the induction step. □

This result can be extended to sets of rational functions.

LEMMA 2.3.12

Let $f_i = p_i/q_i$ be in $F(Q\langle a_1, \ldots, a_r \rangle)$ and suppose $\{f_1, \ldots, f_r\}$ can be computed by a program **P** (with ÷), using t operations. Then there exists $\{v_1, \ldots, v_r\} \subseteq F[Q\langle a_1, \ldots, a_r \rangle]$ such that we can construct a program **P'** (without ÷) using $t \pm$ operations and **P'** computes

$$\{p_1 v_1, q_1 v_1, p_2 v_2, q_2 v_2, \ldots, p_r v_r, q_r v_r\}.$$

Proof. Compute the numerators and denominators separately for each step of the program. □

LEMMA 2.3.13

a-indep $\{p_1 p_2, p_3, \ldots, p_r\} \geq a$-indep $\{p_1, p_3, \ldots, p_r\}$ for any $p_2 \neq 0$.

Sketch of Proof. Let $W = \{w_1, \ldots, w_t\}$. The Interior of $W \triangleq$

$$\{w \mid w = \Sigma \lambda_i w_i, \Sigma \lambda_i = 1, 0 \leq \lambda_i < 1\} \subset W^*,$$

the linear affine subspace generated by W. It is not hard to show (think of the geometry here) that (W − Interior of W) has the same affine dimension as W. Returning to the space defined by $Q\langle a_1, \ldots, a_r \rangle$ let $p_1 X p_2 = \{w_1 w_2 \mid w_1 \in \bar{p}_1, w_2 \in \bar{p}_2\}$ and verify that any point

in $\overline{p_1 X p_2} - \overline{p_1 \times p_2}$ must be in the Interior of $\overline{p_1 \times p_2}$.
It then follows that a-indep $\{p_1 p_2, p_3, \ldots, p_r\} \leq a$-indep $\{p_1 p_2^0, p_3, \ldots, p_r\} = a$-indep $\{p_1, p_3, \ldots, p_r\}$ for any $p_2^0 \in \overline{p_2}$. □

THEOREM 2.3.14

Let $f_1, \ldots, f_r$ be in $F(Q\langle a_1, \ldots, a_r\rangle)$, with $f_i = p_i/q_i$. Then at least $\mathrm{adim}(\{p_1, q_1, \ldots, p_r, q_r\}) \pm$ operations are required to compute $\{f_1, \ldots, f_r\}$ in $F(Q\langle a_1, \ldots, a_r\rangle)$, given $F \cup Q\langle a_1, \ldots, a_r\rangle$.

Proof. Follows from Theorem 2.3.11, Lemma 2.3.12, and repeated application of Lemma 2.3.13. □

2.4 Extending Beyond the Basic Active Substitution Argument

Prior to Strassen's (1973a) deployment of algebraic geometry (see Chapter 5), there were three ways that provided some strengthening of the lower bounds on * complexity. Namely, we can

(i) account for independence amongst the functions $\{f_i\}$ being computed.
(ii) account for nonscalar but nonactive * operations.
(iii) concentrate on special computations; e.g., bilinear forms (see the next section).

In this section, we will describe some of the efforts in (i) and (ii).

LEMMA 2.4.1 (Fiduccia [1971])

Suppose $F \subseteq G$ and $\{f_1, \ldots, f_r\} \subseteq G\langle a_1, \ldots, a_n\rangle$. Let

$$\mathbf{L}_F(\vec{a}, G) \triangleq \left\{ \sum_{i=1}^n c_i a_i + g \,\middle|\, c_i \in F, g \in G \right\};$$

i.e., the set of linear affine functions over F. If $f_1, \ldots, f_r$ are linearly independent over F mod $\mathbf{L}_F(\vec{a}, G)$, then at least r active (relative to F, G) * operations are required to compute $\{f_i\}$.

Proof. Suppose we use only k active operations. Then there exists $\{\alpha_{ij}\} \subseteq F$ and $\{g_i\} \subseteq \mathbf{L}_F(\vec{a}, G)$ such that

$$\alpha_{11}m_1 + \cdots + \alpha_{1k}m_k = f_1 + g_1$$
$$\vdots \qquad \qquad \vdots$$
$$\alpha_{r1}m_1 + \cdots + \alpha_{rk}m_k = f_r + g_r,$$

where the $\{m_i\}$ are the results of the active $*$ operations. But if $k < r$, then the rows of (α) are dependent over F, and hence the $\{f_i\}$ are linearly dependent over F mod $\mathbf{L}_F(\vec{a}, G)$. □

Kirkpatrick (1972b) combines this idea with the basic substitution argument, yielding the following theorem (see also Hopcroft and Kerr [1971]).

THEOREM 2.4.2

Let $\{f_1, \ldots, f_r, f_{r+1}, \ldots, f_s\}$ be computed by a program P using k active $*$ operations, and let $\{f_1, \ldots, f_r\}$ be linearly independent over F mod $\mathbf{L}_F(\vec{a}, G)$. There exists a program P' computing $\{f_{r+1}, \ldots, f_s\}$ from $G \cup \{f_1, \ldots, f_r\} \cup \{a_i\}$ using $k - r$ active $*$ operations.

Proof. Let $m_1, \ldots, m_k$ denote the sequence of active $*$ operations. Then there exists $\{\alpha_{ij}\} \subseteq F$ and $\{g_i\} \subseteq \mathbf{L}_F(\vec{a}, G)$ such that

$$\alpha_{11}m_1 + \cdots + \alpha_{1k}m_k = f_1 + g_1$$
$$\vdots$$
$$\alpha_{r1}m_1 + \cdots + \alpha_{rk}m_k = f_r + g_r.$$

Informally, we want to express $\{m_{k-i} | 0 \leq i \leq r - 1\}$ as linear combinations of the $\{f_j\}$, $\{g_j\}$, and the remaining $\{m_i | 1 \leq i \leq k - r\}$. The linear independence of $\{f_1, \ldots, f_r\}$ guarantees that we are able to "back substitute" for m_k; then m_{k-1}, etc. In a formal proof we have to show that as we "back substitute," the system of equations does not trivialize. □

COROLLARY 2.4.3

At least $p(n + q - 1)$ active $*$ operations are required to compute the entries c_{ij} of $C_{p \times q} = A_{p \times n} B_{n \times q}$, given $G \cup \{b_{ij}\}$, where $G = F(\{a_{ij}\})$. In particular, if $p = q = n$, at least $2n^2 - n$ active operations are required for $C_{n \times n} = A_{n \times n} B_{n \times n}$.

Extending Beyond the Basic Active Subtraction Argument

Proof. Any scheme for the desired computation will require at least $q(p-1)$ more active $*$ operations than to compute $\{c_{ij}\}$, given $G' = G \cup \{b_{ij}\} \cup \{c_{ij} \mid 2 \leq i \leq p, 1 \leq j \leq q\}$. By substituting elements of F for $\{a_{ij} \mid i \geq 2\}$, we observe that the best program for $\{c_{ij}\}$, given G', requires at least as many active $*$ operations as to compute $\{c_{11}, \ldots, c_{1q}\}$, given $F(a_{11}, \ldots, a_{1n}) \cup \{b_{ij}\}$. But from Theorem 2.3.2 we know that nq active operations are required to compute $(a_{11}, \ldots, a_{1n}) B_{n \times q} = (c_{11}, \ldots, c_{1q})$. Hence, at least $(p-1)q + nq = (n+p-1)q$ active $*$ operations are required for the desired computation. □

Fiduccia (1971) provides a matrix-vector formulation for combining independence arguments.

THEOREM 2.4.4

Let the entries of $\Phi_{t \times n}$ and $\vec{\phi}_{t \times 1}$ be in G. We want to compute $\{\psi_i \mid 1 \leq i \leq t\}$ in $G(a_1, \ldots, a_n)$, given $G \cup \{a_i\}$ with $\vec{\psi}_{t \times 1} = \Phi_{t \times n} \vec{a}_{n \times 1} + \vec{\phi}_{t \times 1}$. Suppose Φ has an $r \times s$ submatrix $B_{r \times s}$ with the property: There are no nontrivial vectors $\vec{\alpha}_{1 \times r}$ and $\vec{\beta}_{s \times 1}$ with elements in F such that $\vec{\alpha} B \vec{\beta}$ is in $F \subseteq G$. Then $r + s - 1$ active (relative to F, G) $*$ operations are required to compute $\{\psi_i\}$.

We shall omit the proof and just give one corollary.

COROLLARY 2.4.5 (Winograd [1971] and Munro [1971a])

The computation of the complex product ($\psi_1 = ac - bd$, $\psi_2 = ad + bc$) requires three "active" $*$ operations.

Proof

$$\begin{pmatrix} \psi_1 \\ \psi_2 \end{pmatrix} = B_{2 \times 2} \begin{pmatrix} c \\ d \end{pmatrix} = \begin{pmatrix} a & -b \\ b & a \end{pmatrix} \begin{pmatrix} c \\ d \end{pmatrix}$$

and $B_{2 \times 2}$ satisfies the property expressed in Theorem 2.4.3. The bound is optimal, since the scheme

$$a(c+d) - d(a+b) = ac - bd$$

$$a(c+d) - c(a-b) = ad + bc$$

uses only three multiplications (but five ± operations). However, this method is not numerically stable (e.g., consider $a \gg b$ and $c \gg d$). Note that $\begin{pmatrix} a & b \\ b & a \end{pmatrix}$ does not satisfy the property and, indeed, $\psi_1 = ac + bd$, $\psi_2 = bc + ad$ can be computed over $\mathbf{R}[a,b,c,d]$ in two nonscalar multiplications.

□

As promised at the beginning of this section, we also want to discuss some results that account for certain nonscalar (but not active) $*$ operations. Along these lines, there is a body of results concerning the computation of x^n. We know from the closely related "addition chain" problem (see Knuth [1969]) that for every n, we can compute x^n in $\log n + 0(\log n / \log \log n)$ multiplications. However, lower bounds (as in Erdös [1960] and Schönhage [1973b]) for addition chains cannot be automatically applied, since ± operations can reduce the × complexity of x^n for certain n. On the other hand, growth arguments do apply. Without ÷, it is immediate that $\lceil \log n \rceil$ nonscalar multiplications are required for x^n or any polynomial of degree n. Allowing ÷, we can use Theorem 5.1.14 to show that $\lceil \log n \rceil$ nonscalar $*$ operations are still required. A more direct induction is given by Kung (1973b).

DEFINITION 2.4.6

Let $r(x_1, \ldots, x_l) = p(x_1, \ldots, x_l)/q(x_1, \ldots, x_l)$ be a rational function in $F(x_1, \ldots, x_l)$ with p and q relatively prime polynomials. Define *degree* r = max(degree p, degree q + 1).

THEOREM 2.4.7

Let $r(\vec{x})$ be any rational function of degree n. Then $\lceil \log n \rceil$ nonscalar $*$ operations are required to compute $r(\vec{x})$ in $F(x)$, given $F \cup \{x_i\}$.

Proof. In order to emphasize the nonscalar $*$ complexity, let us express a program as follows (see also Lemma 3.3.2):

$$s_{-l} = x_l$$
$$\vdots$$
$$s_{-1} = x_1$$

Extending Beyond the Basic Active Subtraction Argument

$$s_0 = 1$$

$$s_1 = [\sum_{j=-1}^{0} \alpha_{j,0} s_j] * [\sum_{j=-1}^{0} \beta_{j,0} s_j]$$

$$\vdots$$

$$s_{i+1} = [\sum_{j=-1}^{i} \alpha_{j,i} s_j] * [\sum_{j=-1}^{i} \beta_{j,i} s_j]$$

$$\vdots$$

$$r = \sum_{j=-1}^{k} \alpha_{j,k} s_j$$

with $\{\alpha_{j,i}\}, \{\beta_{j,i}\} \subseteq F$. That is, the s_i denote the results of the non-scalar $*$ operations. We want to prove by induction on k that $\sum_{j=-1}^{k} \gamma_j s_j = r_k(x_1, \ldots, x_l)$ has degree $\leq 2^k$. For notational purposes, let us denote $\sum_{j=-1}^{k} \gamma_j s_j$ as $r_k(\vec{\gamma}, \vec{x}) = p_k(\vec{\gamma}, \vec{x})/q_k(\vec{x})$. The notation implies that $q_k(\vec{x})$ is a fixed polynomial in $F[\vec{x}]$, while $p_k(\vec{\gamma}, \vec{x})$ is a polynomial in $F[\vec{x}]$ whose coefficients depend on the choice of $\vec{\gamma}$.

The basis $k = 0$ is immediate, since each s_j $(-1 \leq j \leq 0)$ has degree ≤ 1. For the induction, we consider two cases ($*$ is $\times$; $*$ is $\div$):

Case 1

$$s_{k+1} = [\sum_{j=-1}^{k} \alpha_{j,k} s_j] \times [\sum_{j=-1}^{k} \beta_{j,k} s_j].$$

Then

$$\sum_{j=-1}^{k+1} \gamma_j s_j = \gamma_{k+1} s_{k+1} + \sum_{j=-1}^{k} \gamma_j s_j$$

$$= \gamma_{k+1}[p_k(\vec{\alpha},\vec{x})/q_k(\vec{x})] \times [p_k(\vec{\beta},\vec{x})/q_k(\vec{x})] + p_k(\vec{\gamma},\vec{x})/q_k(\vec{x})$$

$$= \frac{\gamma_{k+1} p_k(\vec{\alpha},\vec{x}) p_k(\vec{\beta},\vec{x}) + p_k(\vec{\gamma},\vec{x}) q_k(\vec{x})}{q_k(\vec{x}) q_k(\vec{x})}$$

$$\stackrel{\circ}{=} (p_{k+1}(\vec{\gamma},\vec{x}))/q_{k+1}(\vec{x}) \quad \text{(considering } \vec{\alpha} \text{ and } \vec{\beta} \text{ to be fixed).}$$

Case 2

$$s_{k+1} = [\sum_{j=-l}^{k} \alpha_{j,k} s_j] \div [\sum_{j=-l}^{k} \beta_{j,k} s_j].$$

Then

$$\sum_{j=-l}^{k+1} \gamma_j s_j = \gamma_{k+1} s_{k+1} + \sum_{j=-l}^{k} \gamma_j s_j$$

$$= \gamma_{k+1} [p_k(\vec{\alpha}, \vec{x})/q_k(\vec{x})] \times [q_k(\vec{x})/p_k(\vec{\beta},\vec{x})] + p_k(\vec{\gamma},\vec{x})/q_k(\vec{x})$$

$$= \frac{\gamma_{k+1} \; p_k(\vec{\alpha},\vec{x}) q_k(\vec{x}) + p_k(\vec{\gamma},\vec{x}) p_k(\vec{\beta},\vec{x})}{p_k(\vec{\beta},\vec{\gamma}) q_k(\vec{x})}$$

$$\stackrel{\circ}{=} (p_{k+1}(\vec{\gamma},\vec{x})/q_{k+1}(\vec{x})).$$

□

We want to indicate how a "growth" argument can be combined with the active substitution argument to produce lower bounds on the total nonscalar ∗ complexity.

THEOREM 2.4.8

At least $n + \lfloor \log n \rfloor$ ∗ operations are required to compute $\sum_{i=0}^{n} a_i x^i y^{n-i}$

Proof. By induction on u, we can show that $p(\vec{a},x,y) = y^k [\sum_{i=1}^{n} l_i(\vec{a})(x/y)^i + l_0(\vec{a}) + r(x,y)]$ requires at least $\lfloor \log k \rfloor + u$ ∗ operations when there are u linearly independent $\{l_i \mid 1 \leq i \leq n\}$. For the basis ($u = 1$), we note that $p(\vec{a},x,y)$ has degree $\geq (k + 1)$ and hence requires at least $\lceil \log(k + 1) \rceil \leq \lfloor \log k \rfloor + 1$ nonscalar ∗ operations. The induction step follows just as in Theorem 2.1.4, noting that an active operation must exist and that the appropriate substitution effectively changes this to a scalar multiplication. This lower bound can then be applied directly to $\sum_{i=0}^{n} a_i x^i y^{n-i} = y^n \sum_{i=0}^{n} a_i (x/y)^i$; an upper bound of $n + \log n + O(\log n/\log \log n)$ follows from the addition chain results.

□

A more comprehensive development along these same lines can be found in Kedem (1974). Consider the computation of some set of

Extending Beyond the Basic Active Substitution Arguments

functions $\{f_i(x, a_1, \ldots, a_n)\}$ by a program P in $F[x, \vec{a}\,]$, given $F \cup \{x\} \cup \{a_i\}$. Assume $\{f_i(x, \vec{a})\}$ requires at least n active multiplications. Kedem's approach is to carry out symbolically the ith active substitution, introducing a new parameter α_i. After all substitutions have been made, we have a new program P' in $H[x]$, given $H \cup \{x\}$ with $H = F[\alpha_1, \ldots, \alpha_n]$. P' computes a new set of functions $\{f'(x)\}$ as well as some intermediate results; call them $\{A_j(x) \mid 1 \leq j \leq n\}$. If P used m nonscalar (relative to F) multiplications, then P' uses $m - n$ nonscalar (relative to H) multiplications. If we choose $\{\alpha_i\}$ to be elements of F in such a way that $\{f'_i\} \cup \{A_j\}$ requires k nonscalar multiplications, then the original functions $\{f_i(x, a_1, \ldots, a_n)\}$ require at least $n + k$ nonscalar multiplications. To simplify the discussion, we will assume that there is no division (a similar but more complicated result holds when $\div$ is permitted) and that the $\{a_i\}$ are eliminated (substituted for) in the order a_1, $a_2 \cdots$ (otherwise, we can always permute the variable names).

LEMMA 2.4.9

Let P compute $\{f_i(x, a_1, \ldots, a_n)\}$ using m nonscalar multiplications and at least one active (relative to F, $F(x)$) operation. Then there exists a polynomial $A_1(x, a_2, \ldots, a_n)$ in $F_\alpha[x, a_2, \ldots, a_n]$ and a program P' in $F_\alpha[x, a_2, \ldots, a_n]$, given $F_\alpha \cup \{x\} \cup \{a_i\}$ with $F_\alpha = F[\alpha]$ such that P' computes $\{f_i(x, A_1, a_2, \ldots, a_n)\} \cup \{A_1(x, a_2, \ldots, a_n)\}$, using $m - 1$ nonscalar multiplications.

Proof. Let $[q_1(x) + c_1 a_1 + \sum_{i=2}^{n} c_i a_i] \times [q_2(x) + \sum_{i=1}^{n} d_i a_i]$ be the first active operation. To make this a scalar multiplication, we want to set

$$q_1(x) + c_1 a_1 + \sum_{i=2}^{n} c_i a_i = c_1 a_1$$

(postponing the setting of $\alpha_1 \in F$). Renaming a_1, let

$$A_1 = A_1(x, a_2, \ldots, a_n) = \alpha_1 - \frac{1}{c_1} q_1(x) - \frac{1}{c_1} \sum_{i=2}^{n} c_i a_i$$

$$\stackrel{\circ}{=} p_1(x) + \alpha_1 + l_1(a_2, \ldots, a_n).$$

We can view the situation as follows (and hence the lemma):

$$P \longrightarrow p$$

$$\vdots \qquad\qquad \vdots$$

Compute $q_1(x)$ $\qquad\qquad$ compute $q_1(x)$
$\qquad\qquad\qquad\qquad$ compute $p_1(x)$
$\qquad\qquad\qquad\qquad$ compute $A_1 \leftarrow p_1(x) + \alpha + l_1(\vec{a})$

$[q_1(x) + c_1\alpha + \Sigma\, c_i a_i] \times [\ ]$ $\qquad$ $[c_1\alpha] \times [\ ]$

$\qquad\qquad\qquad\qquad \left.\begin{array}{l}\vdots\\ \vdots\end{array}\right\}$ A_1 replaces all occurrences of a_1. $\qquad\qquad\qquad\qquad\qquad\qquad\qquad\qquad\qquad$ □

THEOREM 2.4.10 (Kedem)

Let P compute $\{f_i(x, a_1, \ldots, a_n)\}$ using m nonscalar multiplications and at least n active multiplications. Then there exist polynomials $A_1(x), \ldots, A_n(x)$ in $H[x]$ and a program P' in $H[x]$, given $H \cup \{x\}$ with $H = F[\alpha_1, \ldots, \alpha_n]$ such that P' computes $\{f_i(x, A_1, \ldots, A_n)\} \cup \{A_1(x), \ldots, A_n(x)\}$, using $m - n$ nonscalar operations.

Outline of proof. By repeating the argument in Lemma 2.4.9, we can assert the existence of polynomials $\{p_i\}$ and $\{c_{j,i}\}$ such that the substitutions

$$a_i \leftarrow p_i(x, \alpha_1, \ldots, \alpha_{i-1}) + \alpha_i$$

$$+ \sum_{j=i+1}^{n} c_{j,i}(\alpha_1, \ldots, \alpha_{i-1}) a_j \qquad 1 \leq i \leq n$$

will convert the first n active operations into scalar multiplications. By back substitution we can solve for the $\{a_i\}$ explicitly and obtain a solution of the form

$$A_i = a_i = p_i(x, \alpha_1, \ldots, \alpha_{i-1}) + \alpha_i$$

$$+ \sum_{j=i+1}^{n} [h_{j,i}(\alpha_1, \ldots, \alpha_{j-2}) \cdot (p_j(x, \alpha_1, \ldots, \alpha_{j-1}) + \alpha_j)].$$

Extending Beyond the Basic Active Substitution Arguments

Again we claim that the program P can be transformed into a program P' in $H[x]$ which computes $\{f'_i(x, A_1, \ldots, A_n)\} \cup \{A_j(x)\}$ using n fewer nonscalar operations than P.

□

The surprising observation is that we can infer some complexity results without knowing much about the polynomials p_j and $h_{j,k}$. For example, we have

COROLLARY 2.4.11

Let $p(x, a_1, \ldots, a_n) = \sum_{i=1}^{n} a_i x^{s(i)}$, $s(i)$ an increasing sequence, and let $k = \max_i (s(i+1) - s(i))$. Then $p(x, \vec{a})$ requires $n + \lceil \log k \rceil$ nonscalar multiplications.

Proof. By Theorem 2.1.4, $p(x, \vec{a})$ requires at least n active multiplications. Applying Theorem 2.3.10, we can convert any program P for $p(x, \vec{a})$ to a program P' for $\{A_1(x), \ldots, A_n(x), p(x, A_1(x), \ldots, A_n(x))\}$ which uses n fewer nonscalar multiplications. Let $k = s(m+1) - s(m)$ and set

$$q_1 \stackrel{\circ}{=} \sum_{i=1}^{m} A_i(x) x^{s(i)} \quad \text{and} \quad q_2 \stackrel{\circ}{=} \sum_{i=m+1}^{n} A_i(x) x^{s(i)}.$$

Case 1. $\deg(p(x, \vec{A})) = \deg(q_1 + q_2) \geq s(m+1) \geq s(m+1) - s(m) = k$. By the basic growth argument, $p(x, \vec{A})$ requires at least $\lceil \log k \rceil$ nonscalar multiplications.

Case 2. $\deg(p(x, \vec{A})) < s(m+1)$, which implies $\deg q_1 \geq s(m+1)$. Therefore, for some i_0 $(1 \leq i_0 \leq m)$, $\deg A_{i_0} \geq s(m+1) - s(i_0) \geq s(m+1) - s(m) = k$. Thus, A_{i_0} requires at least $\lceil \log k \rceil$ nonscalar multiplications.

In either case we know that the original program P must have had at least $n + \lceil \log k \rceil$ nonscalar multiplications.

□

The next corollary (2.4.15) has an interesting relation with the problem of computing a polynomial and all its derivatives.

Definition 2.4.12

Let $\vec{\beta} = \langle \beta(1), \ldots, \beta(m) \rangle$ be an increasing sequence of non negative integers and let $\vec{p} = \langle p_1(x), \ldots, p_m(x) \rangle$ be a sequence of polynomials with $p_i(x) = \sum_{j=1}^{m} d_{\beta(j),i} x^{\beta(j)}$. The sequence $\vec{p}$ is called $\vec{\beta}$ *normal* if the determinant of $(d_{\beta(j),i})_{m \times m}$ is nonzero.

Lemma 2.4.13

If $\langle p_1(x), \ldots, p_m(x) \rangle$ is $\vec{\beta}$ normal, then $\{p_i(x)\}$ requires at least m nonscalar multiplications. The proof is similar to Lemma 2.4.1.

Lemma 2.4.14

Let $\langle p_1(x), \ldots, p_m(x) \rangle$ be $\vec{\beta}$ normal and let $f_i(x, \vec{a}) = a_i p_i(x)$ for $1 \leq i \leq m$. Construct P' and $\{A_j \mid 1 \leq j \leq m\}$ as in Theorem 2.4.10, with $f_i(x, A_1, \ldots, A_m) = A_i p_i(x) \stackrel{\circ}{=} \tilde{p}_i(x)$. Then $\langle \tilde{p}_1(x), \ldots, \tilde{p}_m(x) \rangle$ is β normal.

Corollary 2.4.15

The computation of $\{a_1 x, a_2 x^2, \ldots, a_n x^n\}$ requires $2n - 1$ nonscalar multiplications.

Proof. By Theorem 2.3.2, $\{a_i x^i \mid 1 \leq i \leq n\}$ requires n active operations. Moreover, $\langle x, x^2, \ldots, x^n \rangle$ is $\langle 2, 3, \ldots, n \rangle$ normal. So, by Lemmas 2.4.13 and 2.4.14, $\{A_i(x) x^i \mid 1 \leq i \leq n\}$ would require $n - 1$ nonscalar operations, and hence $\{a_i x^i \mid 1 \leq i \leq n\}$ requires $2n - 1$ nonscalar operations.

□

We end this section with a brief discussion of the problem of computing a polynomial and all its normalized derivations. We will also refer to this problem in Chapters 4 and 5. Let $f_0(x, \vec{a}) = f(x, \vec{a}) = \sum_{j=0}^{n} a_j x^j$, $f_i(x, \vec{a}) = f^{(i)}(x, \vec{a})/i!$ and $g_i(x, \vec{a}) = x^i f_i(x, \vec{a})$ for $0 \leq i \leq n$. We are interested in computing $\{f_i(x, \vec{a}) \mid 0 \leq i \leq n\}$. Shaw and Traub (1974) observe that the functions $g_i(x, \vec{a})$ have a very nice structure.

Extending Beyond the Basic Active Substitution Arguments

THEOREM 2.4.16

The functions $g_i(x,\vec{a}) \mid 0 \leq i \leq n$ can be computed in $F[x,\vec{a}]$, given $F \cup \{x\} \cup \{a_i\}$, using $2n-1$ multiplications; furthermore, this bound is optimal.

Proof. By definition,

$$\begin{pmatrix} g_0 \\ g_1 \\ \vdots \\ g_n \end{pmatrix} = \begin{pmatrix} 1 & 1 & 1 & \cdots & & 1 \\ & 1 & 2 & 3 & \cdots & n \\ & & 1 & 3 & 6 & \cdots \\ & & & & & 1 \end{pmatrix} \begin{pmatrix} a_0 \\ a_1 x \\ \vdots \\ a_n x^n \end{pmatrix}$$

$$\stackrel{\circ}{=} (t_{j,i}) \begin{pmatrix} a_0 \\ a_1 x \\ \vdots \\ a_n x^n \end{pmatrix}$$

We claim $t_{j,i} = t_{j-1,i-1} + t_{j,i-1}$. This suggests the following recurrence for computing $g_0, \ldots, g_n$:

$$T_i^{-1} = a_i x^i \qquad 0 \leq i \leq n-1 \quad \Big\} \text{ Initialize}$$
$$T_j^j = a_n x^n \qquad 0 \leq j \leq n$$

$$T_i^j = T_{i-1}^{j-1} + T_{i-1}^j \qquad 0 \leq j \leq n, \; j+1 \leq i \leq n$$

$$g_k = T_n^k \qquad 0 \leq k \leq n$$

Thus, $\{g_k \mid 0 \leq k \leq n\}$ can be computed in $2n-1$ multiplications (i.e., those used in computing $\{a_i x^i \mid 1 \leq i \leq n\}$). Moreover, by inverting the recurrence, we can solve for the $\{a_i x^i \mid 0 \leq i \leq n\}$, given the $\{g_k \mid 0 \leq k \leq n\}$ without using any multiplications; therefore, by Corollary 2.4.14, the computation of $\{g_k\}$ requires $2n-1$ multiplications. □

The results concerning the computation of $\{f_i \mid 0 \leq i \leq n\}$ are not so precise. Using the $\{g_i\}$, we can compute $\{f_i\}$ in $(3n-1) *$ operations while our best lower bound is $n+1$. The recurrence in Theorem 2.4.15 uses $O(n^2)$ ± operations; nevertheless (unlike the results in Chapter 4) this method seems immediately practical. In Problems 4.6 and 4.7, we suggest an $O(n \log n)$ method for computing $\{f_i \mid 0 \leq i \leq n\}$.

2.5 Bilinear Forms and Bilinear Programs

The matrix multiplication problem has provided a focus or goal for arithmetic complexity. The bilinear form of this problem is one characteristic that researchers have attempted to exploit. Moreover, the importance of bilinearity goes beyond matrix multiplication.

DEFINITION 2.5.1

$f(a_1, \ldots, a_p, b_1, \ldots, b_q) \in F[\vec{a}, \vec{b}\,]$ is a *bilinear form* (in $\vec{a}$ and $\vec{b}$) if $f(\vec{a}, \vec{b})$ can be expressed as

$$(a_1, \ldots, a_p)(E)_{p \times q} \begin{pmatrix} b_1 \\ \vdots \\ b_q \end{pmatrix}$$

where $E \in F^{p \times q}$.

Matrix product, complex product, and polynomial product all define sets of bilinear forms.

DEFINITION 2.5.2

A *K bilinear program* has the following format:

$$m_1 = l_1(a_1, \ldots, a_p) \times l'_1(b_1, \ldots, b_q),$$
$$\vdots$$
$$m_n = l_n(a_1, \ldots, a_p) \times l'_n(b_1, \ldots, b_q),$$

$$f_1 = \tilde{l_1}(m_1, \ldots, m_n),$$
$$\vdots$$
$$f_s = \tilde{l_s}(m_1, \ldots, m_n),$$

where the linear combinations are over $K \subseteq F$. (The linear combinations are just a shorthand for the intended $\pm$ operations and multiplications by scalars from K.) For the purposes of this section, we will assume $K = F$, although it is quite reasonable to have $K = \{-1, 0, 1\}$ or $K = Z$ (see Brockett and Dobkin [1973]).

Obviously, a set of bilinear forms can be computed by a bilinear program. But how efficiently? We may be concerned with total operations or we can take the point of view that nonscalar $*$ operations dominate the complexity. This point of view is motivated by the possibility that the variables may represent matrices. (In particular, recall the recursion in Strassen's fast matrix multiplication. Note that Strassen's fast matrix multiplication can be expressed as a bilinear program.) See also the immediately practical application given in Winograd (1973).

This section is primarily concerned with nonscalar $*$ complexity. Assume that we are computing $S = \{f_i(\vec{x})\}$, a set of multivariate polynomials of degree ≤ 2. Winograd (1970a) shows that, with respect to nonscalar multiplications, it is sufficient to consider only programs whose nonscalar multiplications are of the form $l_1(\vec{x}) \times l_2(\vec{x})$. Moreover, an easy modification shows that such programs are asymptotically optimal with respect to total operations.

THEOREM 2.5.3

Let $S = \{f_i(\vec{x}) \mid \deg f_i \leq 2\}$, and let P compute S in $F[\vec{x}]$, given $F \cup \{\vec{x}\}$. Suppose P uses k_1 $\pm$ operations, k_2 scalar and k_3 nonscalar multiplications. Then we can construct an equivalent program P', whose nonscalar multiplications are all of the form $l_1(\vec{x}) \times l_2(\vec{x})$, and P' uses k_3 nonscalar multiplications and less than $9(k_1 + k_2 + k_3)$ total operations.

Proof. Let

$$L_j(p) \overset{\circ}{=} \text{homogeneous part of } p \text{ having degree } j$$

Note: L_0, L_1, L_2, etc., are all linear operations; i.e.,

$$L_j(c_1 p_1 + c_2 p_2) = c_1 L_j(p_1) + c_2 L_j(p_2).$$

For every step s_i of P, we will construct in P' steps $s'_i = (L_0 + L_1 + L_2)(s_i)$ and $s''_i = L_1(s_i)$. Note that deg $s_i \leq 2$ implies $s''_i = s_i$.
(a) If $s_i \leftarrow \lambda s_j$, then $s'_i \leftarrow \lambda s'_j$ and $s''_i \leftarrow \lambda s''_j$.
(b) If $s_i \leftarrow s_j \pm s_k$, then $s'_i \leftarrow s'_j + s'_k$ and $s''_i \leftarrow s''_j \pm s''_k$.
(c) If $s_i \leftarrow s_j \times s_k$, $L_0(s_j) = c_j$, $L_0(s_k) = c_k$) then compute as follows:

$$s'_i \leftarrow (c_j \times s'_k) + (c_k \times s'_j)$$

$$s''_i \leftarrow (c_j \times s''_k) + (c_k \times s''_j)$$

$$+ (s'_j \times s'_k) - c_j c_k$$

P' has $2k_1 + 4k_3 \pm$ operations:
$2k_2 + 4k_3$ scalar $\times$;
k_3 nonscalar $\times$.

$\square$

A similar argument can be used to show that linear programs (i.e., where the only multiplications are scalar multiplications) are asymptotically optimal for computing sets of linear affine functions. Strassen (1973b) and independently Unger extend Winograd's result to show that $\div$ cannot reduce the nonscalar $*$ complexity.

DEFINITION 2.5.4

$F[[x_1, \ldots, x_n]]$ denotes the power series ring over a field F; i.e., the ring of elements of the form

$$\Sigma_{j=0}^\infty \Sigma_{i_1 + \cdots i_n = j} c_{i_1, \ldots, i_n} x_1^{i_1} \cdots x_n^{i_n}$$

with $c_{i_1, \ldots, i_n}$ in F. Any element of the form

$$c_0 + \Sigma_{j=1}^\infty \Sigma_{i_1 + \cdots i_n = j} c_{i_1, \ldots, i_n} x_1^{i_1} \cdots x_n^{i_n}$$

with $c_0 \neq 0$ is invertible in $F[[\vec{x}]]$ and is called a *unit*.

Bilinear Forms and Bilinear Programs

THEOREM 2.5.5

Let $S = \{f_i(\vec{x}) \mid \deg f_i \leq 2\}$ be a set of polynomials. If S can be computed in $F(x_1, \ldots, x_n)$ in k nonscalar $*$ operations, then S can be computed in $F[x_1, \ldots, x_n]$ in k nonscalar multiplications.

Proof. Theorem 2.5.3 simulates a computation in $F[x_1, \ldots, x_n]$ by a computation in $F[\vec{x}] \bmod J$, J being the ideal generated by $\{x_i x_j x_k \mid 1 \leq i \leq j \leq k \leq n\}$; i.e., dropping all terms of degree ≥ 3. We want to first show how to simulate a computation in $F[[x_1, \ldots, x_n]]$, allowing division by units, by a computation in $F[\vec{x}] \bmod J$. Then we want to show how to view a computation in $F(x_1, \ldots, x_n)$ as a computation in $F[[x_1 - \theta_1, \ldots, x_n - \theta_n]]$ for some appropriate choice of $\theta_1, \ldots, \theta_n$.

We can simulate multiplication as before. Now consider the first division (by a unit) which can be considered to be of the form $[c_1 + l_1 + q_1] \div [c_2 - (l_2 + q_2)]$, where c_1, c_2 are constants ($c_2 \neq 0$), l_1 and l_2 are linear, and q_1 and q_2 are quadratic in $\{x_i\}$.

$$[c_1 + l_1 + q_1] \div [c_2 - (l_2 + q_2)]$$

$$= [c_1 + l_1 + q_1] \left[\frac{1}{c_2} + \frac{1}{c_2^2}(l_2 + q_2) + \frac{1}{c_2^3}(l_2 + q_2)^2 + \cdots \right]$$

$$= [c_1 + l_1 + q_1] \left[\frac{1}{c_2} + \frac{1}{c_2^2}(l_2 + q_2) + \frac{1}{c_2^3}(l_2 + q_2)^2 \right] \bmod J$$

$$= \frac{c_1}{c_2} + \frac{c_1}{c_2^2}(l_2 + q_2) + \frac{l_1}{c_2} + \left[\frac{c_1}{c_2^3} l_2 + \frac{1}{c_2^2} l_1 \right] l_2 + \frac{1}{c_2} q_1 \bmod J$$

$$\stackrel{\circ}{=} d_1 + d_2(l_2 + q_2) + \frac{l_1}{c_2} + d_3 q_1 + [d_4 l_2 + d_5 l_1] l_2.$$

That is, in simulating the computation in $F[\vec{x}] \bmod J$, we need use only one nonscalar multiplication to simulate each division.

To complete the proof, we need to show how to view a computation P in $F(x_1, \ldots, x_n)$ as a computation P' in $F[[\tilde{x}_1, \ldots, \tilde{x}_n]]$, $\tilde{x}_i = x_i - \theta_i$, for some choice of $\theta_i \in F$. If P computes $\{f_i(x_1, \ldots, x_n) \mid f_i$ has degree $\leq 2\}$, then P' computes $\{f_i'(\tilde{x}_1, \ldots, \tilde{x}_n) \mid$ where $f_i'(x_1 - \theta_1, \ldots, x_n - \theta_n) = f_i(x_1, \ldots, x_n)$ and hence $\deg f_i' \leq 2\}$.

Consider the first division $p_1(\vec{x}) \div p_2(\vec{x})$. If $p_2(\vec{x})$ has a constant term, then we can view $1/p_2(\vec{x})$ as a power series in $\{x_i\}$. If not, then we want to expand $p_2(\vec{x})$ as a power series in $\{\tilde{x}_i \mid \tilde{x}_i = x_i - \theta_i\}$. For example, $1/(x_1 + x_2) = 1/(1 + (\tilde{x}_1 + x_2)) = 1 - (\tilde{x}_1 + x_2) + (\tilde{x}_1 + x_2)^2 - \cdots$ with $\tilde{x}_1 = x_1 - 1$. To change P to P', choose $\theta_1, \ldots, \theta_n$ so that $\langle \theta_1, \ldots, \theta_n \rangle$ is not a pole or zero of any element (partial result) computed by P. We can view $P' = P|_{x_i \leftarrow \tilde{x}_i + \theta_i}$ as a computation in the power series $F[[\tilde{x}_1, \ldots, \tilde{x}_n]]$ with division only by units. P' can be simulated by a program $P'' = P' \bmod J$, since P' is computing $\{f_i'(\tilde{x}_i) \mid \deg f_i' \leq 2\}$. Finally, we can compute $\{f_i\}$ by the program

$$\begin{cases} \tilde{x}_i = x_i - \theta_i & 1 \leq i \leq n \\ P'' \end{cases}$$

□

Let us now restrict our attention to the nonscalar $*$ complexity required when computing sets $\{f_i(\vec{a}, \vec{b})\}$ of bilinear forms. We know it is sufficient to have all nonscalar $*$ operations be of the form $l_1(\vec{a}, \vec{b}) \times l_2(\vec{a}, \vec{b})$. Moreover, if the $\{a_i\}$ variables do not commute with the $\{b_j\}$ variables, it is sufficient to have $l_1(\vec{a}) \times l_2(\vec{b})$. Winograd (1970a) shows that at worst the nonscalar $*$ complexity doubles when we assume noncommutativity. (Hopcroft and Kerr [1971] show that some saving can be achieved by using commutativity in multiplying $A_{2 \times 2} B_{2 \times n}$.) It should also be apparent that we can organize the computation in three stages:

(1) Compute all necessary linear combinations of the input variables.
(2) Compute all nonscalar multiplications $\{m_i \leftarrow l_{i_1}(\vec{a}) \times l_{i_2}(\vec{b})\}$.
(3) Compute required linear combinations of the $\{m_i\}$.

Summarizing, we have shown that bilinear programs are optimal with respect to nonscalar $*$ complexity and noncommuting $\{\vec{a}\}$ and $\{\vec{b}\}$; they are optimal within a factor of 2 if the $\{\vec{a}\}$ and $\{\vec{b}\}$ commute; and they are asymptotically optimal (i.e., within a factor of some constant c) with respect to total operations.

Bilinear Forms and Bilinear Programs

For the remainder of this section, we will consider only bilinear forms and bilinear programs, with nonscalar multiplications as the measure of complexity. There has been a significant amount of research concerning the structure and symmetries possessed by bilinear programs. We will not be able to discuss all these results, but we should at least mention some of the important papers. By carefully analyzing symmetries, Hopcroft and Kerr (1971) were able to establish a lower bound of $\lceil 7n/2 \rceil$ for $A_{2 \times 2} B_{2 \times n}$ matrix multiplication (assuming only integer scalars are used); repeated application of Algorithm 1.2.1 shows that this bound is optimal. Fiduccia (1972a, b), Gastinel (1971), and Strassen (1972) recognized that the optimal complexity of a set of bilinear forms can be characterized by an algebraic concept, the rank of a tensor. (Indeed, it seems reasonable to infer that the characterization played a significant role in the discovery of Strassen's 2 × 2 method). Fiduccia (1972a, b) discussed the structure of bilinear programs and the scope of possible applications (including the idea of computationally dual problems). Hopcroft and Musinski (1973), and independently Probert (1973), exploit duality in a particularly interesting application; namely, they show that $A_{n_1 \times n_2} B_{n_2 \times n_3}$ and $A_{n_{\pi(1)} \times n_{\pi(2)}} B_{n_{\pi(2)} \times n_{\pi(3)}}$ have the same complexity (for any permutation π). A theory of bilinear forms and programs can be found in Brockett and Dobkin (1973) and Dobkin (1973).

In presenting some of these ideas, we will follow the exposition of Brockett and Dobkin (1973). Let $\vec{a} = (a_1, \ldots, a_p)^T$, $\vec{b} = (b_1, \ldots, b_q)^T$ and let $\langle\langle \vec{x}, \vec{y} \rangle\rangle$ represent the inner product $\Sigma x_i y_i$. A bilinear form $f_i(\vec{a}, \vec{b})$ can be defined by a matrix G_i satisfying $f_i(\vec{a}, \vec{b}) = \langle\langle \vec{a}, G_i \vec{b} \rangle\rangle$. A set of forms $S = \{f_i \mid 1 \leq i \leq m\}$ can be defined by the matrix $G(\vec{s}) \triangleq \sum_{i=1}^{m} s_i G_i$, by the trilinear form

$$H(\vec{s}, \vec{a}, \vec{b}) \triangleq \sum_{i=1}^{m} s_i f_i = \sum_{i=1}^{m} s_i \langle\langle \vec{a}, G_i \vec{b} \rangle\rangle,$$

or by the third-order tensor $(h_{ijk})_{m \times p \times q}$, where $H(\vec{s}, \vec{a}, \vec{b}) = \sum_i \sum_j \sum_k h_{ijk} s_i a_j b_k$. The $\{s_i\}$ are new indeterminates, used to delineate the form being computed; their introduction simplifies the discussion.

In the form $H(\vec{s},\vec{a},\vec{b})$, each type of variable plays a distinct role; call $\vec{s}$ the function separators, $\vec{a}$ the left-hand indeterminates, $\vec{b}$ the right-hand indeterminates. If we interchange (permute) the role of the variables, we are defining a new set of bilinear forms. Choosing one permutation, let $\hat{H}(\vec{b},\vec{s},\vec{a})$ be $H(\vec{s},\vec{a},\vec{b})$, except that now $\vec{b}$ is considered to be the function separators, $\vec{s}$ the left-hand indeterminates, and $\vec{a}$ the right-hand indeterminates. For example, $H(\vec{s},\vec{a},\vec{b}) \triangleq (a_1 b_1 - a_2 b_2)s_1 + (a_1 b_2 + a_2 b_1)s_2$ defines the real and imaginary parts of a complex product, while $\hat{H}(\vec{b},\vec{s},\vec{a})$ defines the forms $s_1 a_1 + s_2 a_2$ and $s_2 a_1 - s_1 a_2$. By convention $\hat{H}(\vec{b},\vec{s},\vec{a})$, $\hat{G}(\vec{b})$, and $\{\hat{G}_k \mid 1 \leq k \leq q\}$ all define the same forms.

THEOREM 2.5.6

Suppose S_1, defined by $H(\vec{s},\vec{a},\vec{b})$, is computable in n (nonscalar multiplication) steps. Then S, defined by $\hat{H}(\vec{b},\vec{s},\vec{a})$, is also computable in n steps. In fact, any permutation of the variables (e.g., $\hat{H}(\vec{a},\vec{b},\vec{s})$, etc.) will lead to another computationally equivalent set of bilinear forms.

Proof. Let the multiplication steps be denoted by $m_l \triangleq \langle\langle \vec{v}^l, \vec{a} \rangle\rangle \cdot \langle\langle \vec{w}^l, \vec{b} \rangle\rangle$, $1 \leq l \leq n$; i.e., $m_l = (\Sigma v_i^l a_i) \times (\Sigma w_j^l b_j)$. Recalling Definition 2.5.2, it follows that $f_i(\vec{a},\vec{b}) = \sum_{l=1}^{m} u_i^l \langle\langle \vec{v}^l, \vec{a} \rangle\rangle \cdot \langle\langle \vec{w}^l, \vec{b} \rangle\rangle$ for some appropriate $u_i^1, \ldots, u_i^m$. Thus,

$$H(\vec{s},\vec{a},\vec{b}) = \sum_{i=1}^{m} s_i \langle\langle \vec{a}, G_i \vec{b} \rangle\rangle$$

$$= \sum_{i=1}^{m} s_i \sum_{l=1}^{n} u_i^l \langle\langle \vec{v}^l, \vec{a} \rangle\rangle \cdot \langle\langle \vec{w}^l, \vec{b} \rangle\rangle$$

$$= \sum_{l=1}^{n} \sum_{i=1}^{m} u_i^l s_i \langle\langle \vec{v}^l, \vec{a} \rangle\rangle \cdot \langle\langle \vec{w}^l, \vec{b} \rangle\rangle$$

$$= \sum_{l=1}^{n} \langle\langle \vec{u}^l, \vec{s} \rangle\rangle \cdot \langle\langle \vec{v}^l, \vec{a} \rangle\rangle \cdot \langle\langle \vec{w}^l, \vec{b} \rangle\rangle.$$

Now permute the variables as indicated by $\hat{H}$:

Bilinear Forms and Bilinear Programs 41

$$\hat{H}(\vec{b},\vec{s},\vec{a}) = \sum_{l=1}^{n} \langle\langle \vec{w}^l, \vec{b}\rangle\rangle \cdot \langle\langle \vec{u}^l, \vec{s}\rangle\rangle \cdot \langle\langle \vec{v}^l, \vec{a}\rangle\rangle$$

$$= \sum_{l=1}^{n} \sum_{k=1}^{q} w_k^l b_k \langle\langle \vec{u}^l, \vec{s}\rangle\rangle \cdot \langle\langle \vec{v}^l, \vec{a}\rangle\rangle$$

$$= \sum_{k=1}^{q} b_k \sum_{l=1}^{n} w_k^l \langle\langle \vec{u}^l, \vec{s}\rangle\rangle \cdot \langle\langle \vec{v}^l, \vec{a}\rangle\rangle.$$

This indicates how to compute $\hat{S}$ in n steps.

□

Returning to the complex product example, the following program computes $H(\vec{s}, \vec{a}, \vec{b})$:

$$m_1 = (a_1 + a_2)b_1 \qquad \langle\langle \vec{a}, G_1\vec{b}\rangle\rangle = m_1 + m_2$$

$$m_2 = -a_2(b_1 + b_2) \qquad \langle\langle \vec{a}, G_2\vec{b}\rangle\rangle = m_1 + m_3$$

$$m_3 = a_1(-b_1 + b_2)$$

Thus,

$$u^1 = (1,1)^T, \quad u^2 = (1,0)^T, \quad u^3 = (0,1)^T$$

$$v^1 = (1,1)^T, \quad v^2 = (0,-1)^T, \quad v^3 = (1,0)^T$$

$$w^1 = (1,0)^T, \quad w^2 = (1,1)^T, \quad w^3 = (-1,1)^T$$

We can then compute $\hat{H}(\vec{b},\vec{s},\vec{a})$ by the program

$$m_1 = (s_1 + s_1)(a_1 + a_2) \qquad \langle\langle \vec{s}, \hat{G}_1\vec{a}\rangle\rangle = \hat{m}_1 + \hat{m}_2 - \hat{m}_3$$

$$m_2 = s_1(-a_2) \qquad \langle\langle \vec{s}, \hat{G}_2\vec{a}\rangle\rangle = \hat{m}_2 + \hat{m}_3$$

$$m_3 = s_2(a_1)$$

In the same way, we can define $H(\vec{s},\vec{a},\vec{b})$ for the bilinear forms associated with $A_{m \times n} B_{n \times p}$ and verify that permuting the variables in H defines the bilinear forms associated with $S_{p \times m} A_{m \times n}$, etc.

Note that additions are not in general preserved by the indicated program transformation; e. g., $A_{n \times n} B_{n \times 1}$ requires $n(n-1) \pm$ operations, whereas $S_{n \times 1} B_{1 \times n}$ does not require any $\pm$ operations. One more example deserves mention. Let $H(\vec{s}, \vec{a}, \vec{b})$ define the forms associated with (deg $p-1$ by deg $q-1$) polynomial multiplication; i.e.,

$$G(\vec{s}) = \begin{pmatrix} s_1 & \cdots & s_2 & \cdots & s_q \\ s_2 & \cdots & s_3 & \cdots & s_{q+1} \\ s_p & \cdots & & \cdots & s_{p+q-1} \end{pmatrix}$$

Then $\hat{H}(\vec{b}, \vec{s}, \vec{a})$ defines the forms associated with $S_{q \times p} \vec{a}$, where S is a Toeplitz matrix. (A Toeplitz matrix $T = (t_{i,j})$ satisfies $t_{i,j} = t_{i+1, j+1}$).

To complete this section, we want to establish the relation between the optimal complexity for H and the rank of the tensor (h_{ijk}).

DEFINITION 2.5.7

(a) The *rank* of a tensor $H = (h_{i,j})$ of order 2 (i.e., a matrix) is the least n such that there exist $\vec{e}^1, \vec{e}^2, \ldots, \vec{e}^n, \vec{f}^1, \ldots, \vec{f}^n$ satisfying $h_{i,j} = \sum_{l=1}^{n} e_i^l f_j^l$; equivalently, if E is the matrix whose columns are the $\vec{e}^l$, and F the matrix whose rows are the $\vec{f}^l$, then $H = EF$.

(b) The *rank* of a tensor $H = (h_{i,j,k})$ of order 3 is the least n such that there exist $\vec{e}^l, \vec{f}^l, \vec{g}^l$ ($1 \leq l \leq n$) satisfying $h_{i,j,k} = \sum_{l=1}^{n} e_i^l f_j^l g_k^l$.

This definition of rank easily generalizes to tensors of any order. For a matrix, the definition is equivalent to the more operational definitions based on row or column independence. Unfortunately, there is no such operational equivalent for tensors of order 3.

Bilinear Forms and Bilinear Programs

THEOREM 2.5.8

Let $\{G_i \mid 1 \leq i \leq m\}$, $G(\vec{s})$, and $(h_{ijk})_{m \times p \times q}$ all represent the same set S of bilinear forms. The following statements are equivalent:

(a) S can be computed in n steps.

(b) There exist $\vec{u}^1, \ldots, \vec{u}^n$ and rank 1 matrices $D_1, \ldots, D_n$ such that
$$G_i = \sum_{l=1}^{n} u_i^l D_l \quad (1 \leq i \leq m);$$
thus $G(\vec{s}) = \sum_{l=1}^{n} (\Sigma\, u_i^l s_i) D_l$.

(c) (h_{ijk}) has rank $\leq n$.

Sketch of Proof. The proof follows mostly from the definitions. For example, consider (a) implies (b):

$$\langle\!\langle \vec{a}, G_i \vec{b} \rangle\!\rangle = \sum_{l=1}^{n} u_i^l \langle\!\langle \vec{v}^l, \vec{a} \rangle\!\rangle \cdot \langle\!\langle \vec{w}^l, \vec{b} \rangle\!\rangle$$

$$= \sum_{l=1}^{n} u_i^l \langle\!\langle (\vec{v}^l)^T \vec{a}, (\vec{w}^l)^T \vec{b} \rangle\!\rangle$$

$$= \sum_{l=1}^{n} u_i^l \langle\!\langle \vec{a}, \vec{v}^l (\vec{w}^l)^T \vec{b} \rangle\!\rangle$$

$$= \sum_{l=1}^{n} u_i \langle\!\langle \vec{a}, D_l \vec{b} \rangle\!\rangle.$$

□

Thus, the optimal complexity can be characterized in terms of a minimal spanning of $G(\vec{s})$ by rank 1 matrices. For example (see and compare the presentations of Brockett and Dobkin [1973], Fiduccia [1972a, b] and Gastinel [1971]),

$$G(\vec{s}) = \begin{pmatrix} s_1 & 0 & s_2 & 0 \\ 0 & s_1 & 0 & s_2 \\ s_3 & 0 & s_4 & 0 \\ 0 & s_3 & 0 & s_4 \end{pmatrix}$$

defines the forms associated with 2 by 2 matrix multiplication; $G(\vec{s})$ can be represented (according to Strassen) as

$$CA(\vec{s})B \quad \text{or} \quad CBA(\vec{s}) = \sum_{l=1}^{7} (\text{column } l \text{ of } C)(\text{row } l \text{ of } B)A(s)$$

$$\stackrel{\circ}{=} \Sigma D_l A(s)$$

with

$$C = \begin{pmatrix} 1 & -1 & 1 & 0 & 0 & 0 & -1 \\ 0 & 1 & 0 & 0 & 0 & 0 & 0 \\ 0 & 0 & 0 & 1 & 0 & 0 & 1 \\ -1 & 0 & 0 & -1 & 1 & -1 & 0 \end{pmatrix}$$

$$B = \begin{pmatrix} 1 & 0 & 0 & -1 \\ 0 & 0 & 0 & 1 \\ 0 & 0 & 1 & 1 \\ 1 & 0 & 0 & 0 \\ 1 & 1 & 0 & 0 \\ 0 & 1 & 0 & 1 \\ 1 & 0 & 1 & 0 \end{pmatrix}$$

$$A(\vec{s}) = \begin{pmatrix} s_1 + s_4 & & & & & & \\ & s_2 - s_1 & & & & & \\ & & s_2 + s_4 & & & & \\ & & & s_3 - s_4 & & & \\ & & & & s_3 + s_1 & & \\ & & & & & s_1 & \\ & & & & & & s_4 \end{pmatrix}$$

Fiduccia (1972a) suggests some interesting heuristics for constructing such a decomposition. But the fact remains that even for "small" matrices we do not know how to obtain a minimal decomposition. Consider the forms associated with $A_{3 \times 3} B_{3 \times 3}$; the best method known so far requires 24 steps, a 21-step method would provide the basis for an n by n matrix multiplication algorithm with complexity $O(n^{\log_3 21}) \approx O(n^{2.77})$.

2.6 Problems Related to Matrix Multiplication

In Chapter 1 we introduced the surprising result of Strassen (1969) that n by n matrices can be multiplied in $O(n^{\log 7})$ arithmetic operations. In the previous sections of this chapter we concentrated on showing lower bounds on the number of arithmetics required in computing polynomials and multiplying matrices. In this section we will use "fast matrix multiplication" to provide upper bounds on the computation of certain polynomial and matrix problems. It should be noted that matrix multiplication provides (a basis for) asymptotically fast algorithms for a number of important "nonnumeric" problems such as computing the transitive closure of a directed graph (Munro [1971b]) and parsing context free languages (Valiant [1974]). Such applications help link algebraic complexity with other aspects of computational complexity.

Before discussing the numeric applications, we wish to comment on the optimality of 2 by 2 matrix multiplication. Winograd presents the following algorithm, which uses 7 (noncommutative) multiplications and 15 (rather than 18, as in Strassen [1969]) ± operations. Let

$$A = \begin{pmatrix} a_{11} & a_{12} \\ a_{21} & a_{22} \end{pmatrix} \quad B = \begin{pmatrix} b_{11} & b_{12} \\ b_{21} & b_{22} \end{pmatrix}$$

and compute

$$\begin{aligned} S_1 &= a_{21} + a_{22} & S_5 &= b_{12} - b_{11} \\ S_2 &= S_1 - a_{11} & S_6 &= b_{22} - S_5 \\ S_3 &= a_{11} - a_{21} & S_7 &= b_{22} - b_{12} \\ S_4 &= a_{12} - S_2 & S_8 &= S_6 - b_{21} \end{aligned}$$

$$P_1 = S_2 S_6 \qquad P_5 = S_1 S_5$$

$$P_2 = a_{11} b_{11} \qquad P_6 = S_4 b_{22}$$

$$P_3 = a_{12} b_{21} \qquad P_7 = a_{22} S_8$$

$$P_4 = S_3 S_7$$

$$S_9 = P_3 + P_2 \qquad S_{13} = S_{12} + P_6$$

$$S_{10} = P_1 + P_2 \qquad S_{14} = S_{11} - P_7$$

$$S_{11} = S_{10} + P_4 \qquad S_{15} = S_{11} + P_5$$

$$S_{12} = S_{10} + P_5$$

Then

$$AB = \begin{pmatrix} S_9 & S_{13} \\ S_{14} & S_{15} \end{pmatrix}$$

Using Strassen's (1969) analysis of the number of arithmetics required to multiply n by n matrices based on judicious application of his 18 addition 2 by 2 scheme, it can be shown that this scheme yields an algorithm using at most $4.54\, n^{\log 7}$ arithmetic operations (rather than $4.7\, n^{\log 7}$). It may seem curious that a reduction from 25 to 22 arithmetics (8%) in the basic scheme should produce only a 3% drop in the general case. We observe, however, that a large proportion of the "work" is in multiplying fairly small (8 by 8 or so) matrices in the "standard" manner. An analysis by Fischer (1974) has shown that a careful application of Winograd's form will produce a $3.92\, n^{\log 7}$ algorithm and, of course, that Strassen's original method can produce a slightly more costly scheme.

From the analysis of Strassen's algorithm in Section 1.2, we observe that, given an m multiplication algorithm for multiplying k by k matrices over an arbitrary (i.e., noncommutative) ring, we can achieve an $0(n^{\log_k m})$ algorithm for multiplying n by n matrices. It is natural to ask whether or not we can improve on Winograd's method for 2 by 2 matrices to reduce further the arithmetics required in general. This is not the case. Hopcroft and Kerr (1971) and Wino-

Problems Related to Matrix Multiplication

grad (1971) have shown that 7 multiplications are required to multiply 2 by 2 matrices, and Probert (1973) has shown that 15 additions are required for any 7 multiplication algorithm that is performed over an arbitrary ring. Any further advances in "fast matrix multiplication" must, then, come from studying larger systems. We now turn from the problem of matrix multiplication to some of its applications.

In Sections 2.1 and 2.2 it was shown that Horner's rule is an optimal method for evaluating a general polynomial, given the coefficients $\{a_i\}$ and the indeterminate x. But suppose we are given the coefficients of the polynomial and a large number of points at which it is to be computed. How many arithmetics are required per point in the limiting case? Is repeated application of Horner's rule optimal? We will see more detailed answers to this and related questions in Chapters 3 and 4, but the optimality of Horner's rule for multipoint evaluation is simply denied by the following arguments.

LEMMA 2.6.1

We can compute $\{p_i(\vec{a}_i, x_k) \mid p_i(\vec{a}_i, x_k) = \sum_{j=0}^{n} a_{ij} x_k^j \text{ for } 1 \leq i \leq n, 1 \leq k \leq n\}$, given $F \cup \{a_{ij}\} \cup \{x_k\}$, using $O(n^{\log_2 7})$ arithmetic operations.

Proof. Define the n by n matrices A and X as

$$A = \begin{pmatrix} a_{1,1} & a_{1,2} & \cdots & a_{1,n} \\ a_{2,1} & & & \\ \vdots & & & \vdots \\ a_{n,1} & \cdots & \cdots & a_{n,n} \end{pmatrix}$$

and

$$X = \begin{pmatrix} x_1 & x_2 & \cdots & x_n \\ x_1^2 & & & \\ \vdots & & & \vdots \\ x_1^n & & & x_n^n \end{pmatrix}$$

Now compute the matrix Y as $Y = \{y_{i,k}\} = AX$ and so compute

$$p_i(x_k) = y_{i,k} + a_{i,0}$$

We observe that X can be determined in $n^2 - n$ multiplications and that n^2 additions are required to form the $y_{i,k} + a_{i,0}$ sums. The computation time is dominated by the $O(n^{\log 7})$ arithmetics used to compute the matrix product AX.

□

LEMMA 2.6.2

A polynomial of degree n may be evaluated at $\sqrt{n}$ points in $O(n^{\log 7)/2})$ arithmetic operations.

Proof. The trick is again to embed the bulk of the computation in a matrix multiplication. Assume, without loss of generality, that n is a perfect square, and define the $\sqrt{n}$ by $\sqrt{n}$ matrices A and X as follows:

$$A = \begin{pmatrix} a_1 & a_2 & \cdots & a_{\sqrt{n}} \\ a_{\sqrt{n}+1} & \cdots & \cdots & a_{2\sqrt{n}} \\ \vdots & & & \\ a_{n-\sqrt{n}+1} & \cdots & \cdots & a_n \end{pmatrix}$$

and

$$X = \begin{pmatrix} x_1 & x_2 & \cdots & x_{\sqrt{n}} \\ x_1^2 & & & \\ & & & \vdots \\ \vdots & & & \\ x_1^{\sqrt{n}} & \cdots & \cdots & x_{\sqrt{n}}^{\sqrt{n}} \end{pmatrix}$$

Problems Related to Matrix Multiplication   49

Now compute $Y = \{y_{i,k}\} = AX$. Since $y_{i,k} = \sum_{j=1}^{\sqrt{n}} a_{(i-1)\sqrt{n}+j} x_k^j$, we may apply Horner's rule to find

$$p(x_k) = \sum_{j=0}^{n} a_j x_k^j = \sum_{j=1}^{\sqrt{n}} y_{j,k} (x_k^{\sqrt{n}})^{j-1} + a_0 \qquad \text{for } k = 1, \ldots, \sqrt{n}.$$

The determination of X and final step in the evaluation of $p(x_k)$ together require $2(n - \sqrt{n})$ multiplications and $n - \sqrt{n}$ additions. The computation time is dominated by the multiplication of two $\sqrt{n}$ by $\sqrt{n}$ matrices. This can be accomplished in $O(n^{(\log 7)/2})$ arithmetic operations.

Let us write $f(n) = o(f(n))$ to indicate that $\lim_{n \to \infty} g(n)/f(n) = 0$.

THEOREM 2.6.3

A polynomial of degree n may be evaluated at m points $(m > n)$ in $O(mn^{.91}) = o(n) \cdot m$ arithmetic operations.

Proof. The result follows by repeated application of Lemma 2.6.2.
□

At this point it seems appropriate to reiterate our remarks concerning "upper bounds" and "practical algorithms." The techniques suggested above were certainly not chosen for their practicality but rather because they suggest the possibility of better methods.

Let us now consider the problem of matrix inversion. Strassen (1969) suggested a way (essentially 2 by 2 Gaussian elimination) in which inversion may be reduced to (fast) matrix multiplication. Bunch and Hopcroft (1974) describe Strassen's suggestion as the following block (LDU) factorization. Let

$$A = \begin{pmatrix} a_{11} & a_{12} \\ a_{21} & a_{22} \end{pmatrix} = \begin{pmatrix} I & 0 \\ a_{21} a_{11}^{-1} & I \end{pmatrix} \begin{pmatrix} a_{11} & 0 \\ 0 & d \end{pmatrix} \begin{pmatrix} I & a_{11}^{-1} a_{12} \\ 0 & I \end{pmatrix}$$

where $d = a_{22} - a_{21} a_{11}^{-1} a_{12}$.

Therefore, we compute A^{-1} as

$$A^{-1} = \begin{pmatrix} I & -a_{11}^{-1}a_{12} \\ 0 & I \end{pmatrix} \begin{pmatrix} a_{11}^{-1} & 0 \\ 0 & d^{-1} \end{pmatrix} \begin{pmatrix} I & 0 \\ -a_{21}a_{11}^{-1} & I \end{pmatrix}$$

The technique uses two inversions, six multiplications and two additions. Since it does not make use of the commutativity of multiplication, we may apply it recursively. Assuming no singular submatrices (a_{11} or d) are encountered and that the matrix multiplication algorithm used requires $M(n)$ arithmetics, the run time of this algorithm $I(n)$ may be analyzed as

$$I(n) = 2I(n/2) + 6M(n/2) + n^2/2$$

$$= 4I(n/4) + 12M(n/4) + n^2/4 + 6M(n/2) + n^2/2$$

$$\leq 3 \sum_{i=1}^{\lceil \log n \rceil} 2^i M(n/2^i) + 0(n^2).$$

Using an $0(n^{\log 7})$ multiplication algorithm, this produces an $0(n^{\log 7})$ inversion scheme; a more careful application of the Winograd scheme results in a $4.8n^{\log 7}$ algorithm. It is also observed that if $M(2n) \geq (2+\epsilon)M(n)$ for all n (e.g., if $M(n) = cn^\alpha$ or $cn^\alpha \log n^\beta$ ($\alpha \geq 2$)), then $I(n) = 0(M(n))$.

The algorithm, however, requires that a number of smaller matrices be inverted to produce intermediate results. Some of these matrices could be singular even though the original matrix is not. In such a case the algorithm as stated would fail. While we may argue that the set of cases in which this happens is of measure zero (if we are working over the reals), the problem can be avoided. Bunch and Hopcroft (1974) show how to perform an appropriate permutation of the given matrix to avoid this problem. If, however, we are working over **R** or **C**, a suggestion of Schönhage (1973a) provides a very neat solution (although it requires more arithmetics than the Bunch-Hopcroft technique and is numerically less stable). By A^* let us denote the conjugate transpose of A (the $\langle i,j \rangle$ element of A^* is the complex conjugate of a_{ji}). Now observe that $A^{-1} = (A^*A)^{-1}A^*$, and that A^*A is a positive-definite Hermitian matrix. This guarantees that a_{11} and $d = a_{22} - a_{21}a_{11}^{-1}a_{12}$ are not only nonsingular, but also positive-definite Hermitian. In other words, assuming A is non-

Problems Related to Matrix Multiplication

singular, Strassen's algorithm for matrix inversion applied to A^*A will encounter no difficulties with singular matrices.

We have now essentially reduced matrix inversion to multiplication. The reduction may also be made in the opposite direction, as observed by Munro (1973). An n by n product may be found by inverting a $3n$ by $3n$ matrix, as indicated by the identity

$$\begin{pmatrix} I & A & 0 \\ 0 & I & B \\ 0 & 0 & I \end{pmatrix}^{-1} = \begin{pmatrix} I & -A & AB \\ 0 & I & -B \\ 0 & 0 & I \end{pmatrix}$$

If $I(n)$ denotes the number of arithmetics required to invert an n by n matrix, this implies $I(3n) \geqslant M(n)$. By observing that $M([n/4]) \geqslant kM(n)$ for some $k > 0$, it can be verified that $I(n) \geqslant kM(n)$. We can summarize our investigation of matrix inversion in a theorem.

THEOREM 2.6.4

Subject to the assumption that the number of arithmetics required for n by n matrix multiplication grows "uniformly in n" (i.e., $(M(2n) \geqslant (2 + \epsilon)M(n)$ for some $\epsilon > 0$ and all n), the number of arithmetics necessary for n by n matrix inversion and matrix multiplication are of the same order of magnitude.

Finally, we note that, based on the identity $\det A = \det a_{11} \cdot \det (a_{22} - a_{21} a_{11}^{-1} a_{12})$ we can find the determinant of a matrix with arithmetic complexity proportional to that required to perform matrix inversion.

PROBLEMS

2.1 Prove that $\lceil n/2 \rceil$ nonscalar $*$ operations are required to compute $\sum_{i=1}^{n} a_i^2$ in $C(a_1, \ldots, a_n)$ given $C \cup \{a_i\}$.

2.2 Let $\psi_{1 \times t} = (a_1, \ldots, a_n) \Phi_{n \times t} + \phi_{1 \times t}$ with the entries of Φ and ϕ in G. Suppose there are rows $\Phi_1, \ldots, \Phi_u$ of Φ which are linearly independent with respect to some $F \subseteq G$. Show that u active (relative to F, G) $*$ operations are required to compute $\psi_1, \ldots, \psi_t$.

2.3 Let $t = (n + q - 1)p$. Show that at least $t\pm$ operations and $t *$ operations are required to compute the elements c_{ij} of $C_{p \times q} = A_{p \times n} B_{n \times q}$.

2.4 Prove that at least $2(n + m)$ total arithmetic operations are required to compute $\{ \sum_{j=0}^{n} a_j x_k^j \mid 1 \leq k \leq m \}$.

2.5 Show how to compute $\sum_{i=0}^{n} a_i x^i$ and $\sum_{i=1}^{n} i a_i x^{i-1}$ in

(a) n active $*$ operations;

(b) $n + 2\sqrt{n}$ total $*$ operations.

2.6 Show how to compute $\sum_{i=0}^{n} a_i x_1^i$ and $\sum_{i=0}^{n} a_i x_2^i$ using less than $2n *$ operations. *Hint*: Consider Fiduccia's (1972a) method for computing $A_{n \times n} B_{n \times 2}$.

2.7 (Winograd [1970a].) Show that for the lower bounds based on linear independence (e.g., Theorems 2.1.4, 2.2.4, 2.3.2, 2.3.4) the assumption that F is infinite is not necessary. For example, in Theorem 2.1.4, if F is finite, then consider the computation to be in $F'(\vec{a}, x)$, given $F'(x) \cup \{a_i\}$ where $F' = F(z)$, z a new indeterminate. If the l_i are linearly independent over F, they will also be linearly independent over F'.

2.8 (Sieveking [1972] and Strassen [1973b].) Extend Theorem 2.5.5 as follows: Suppose we can compute $\{f_i(x_1, \ldots, x_r)\}$ in $F(x_1, \ldots, x_r)$ in k nonscalar $*$ operations, and suppose each polynomial f_i has degree $\leq n$. Then we can compute $\{f_i(x_1, \ldots, x_r)\}$ in $F[x_1, \ldots, x_r]$ in $\leq cnk$ nonscalar multiplications for some constant c. For $r = 1$, see the proof of Theorem 4.4.1.

2.9 Show that linear programs are asymptotically optimal for computing sets of linear functions.

2.10 (Kirkpatrick [1972a].) In computing in $F[\vec{a}, x]$ show that

(a) $2n-1 \pm$ operations are required for $\sum_{i=0}^{n} a_i x^i$ and $\sum_{i=1}^{n} i a_i x^{i-1}$;

(b) $2n \pm$ operations are required for $\sum_{i=0}^{n} a_i x_1^i$ and $\sum_{i=0}^{n} a_i x_2^i$.

2.11 (Strassen [1972].) Let f_n be the continued fraction defined by $f_1 = 1/x_1$ and $f_{j+1} = 1/(x_{j+1} + f_j)$ for $1 \leq j \leq n - 1$. Show that $n *$ operations are required to compute f_n in $F(x_1, \ldots, x_n)$, given $F \cup \{x_1, \ldots, x_n\}$.

CONJECTURES–OPEN PROBLEMS

1. For all k, show that there exists n such that $\sum_{i=0}^{n} a_i x^i$ and $\sum_{i=1}^{n} i a_i x^{i-1}$ (or $\sum_{i=0}^{n} a_i x_1^i$ and $\sum_{i=0}^{n} a_i x_2^i$) requires more than $n + k$ nonscalar $*$ operations.

2. Does there exist a set of linear functions $\{f_i\}$ whose computation cannot be optimally realized (with respect to total arithmetic operations) by a linear program? See also Morgenstern (1973b).

3. For all c, there exists a polynomial $f(x_1, \ldots, x_r)$ which can be computed in $F(x_1, \ldots, x_r)$, given $F \cup \{x_i\}$, using k (nonscalar) $*$ operations, but whose computation in $F[x_1, \ldots, x_r]$ cannot be realized in ck (nonscalar) multiplications. The polynomial $f(\vec{x}) = \det(x_{ij})$ is a good candidate.

4. Prove that $\sum_{i=0}^{n} x^i$ requires $O(\log n)$ $\pm$ operations when computed in $F[x]$.

5. Can we possibly perform $n \times n$ matrix multiplication in $O(n^2)$ nonscalar multiplications? Optimists may find some encouragement in the recent result by Brockett and Dobkin, which shows that for $p \leq \log n$, $n \times p$ by $p \times n$ matrix multiplication can be performed in $n^2 + O(n^2)$ nonscalar multiplications.

Chapter 3

LOWER BOUNDS AND CONCEPTS RELATED TO ALGEBRAIC INDEPENDENCE

This chapter is motivated by the computation of specific polynomials. For example, the fact that the evaluation of $\sum_{i=0}^{n} a_i x^i$ from $\{a_0, \ldots, a_n, x\}$ requires n additions does not imply that computing $\sum_{i=0}^{n} x^i$ requires the same complexity. Formally, we are concerned with computing $p(x) \in F[x]$ in $F(x)$, given $F \cup \{x\}$. Given an arbitrary but fixed polynomial, can we improve upon Horner's rule?

3.1 Preconditioning

Consider $p(x) \in F[x]$, where F is the field of real or complex numbers. Is the representation of $p(x)$ by its coefficients, the most appropriate one with regards to its evaluation (at possibly a large number of points)? It is clear we could also represent a polynomial of degree n by its values at any $n + 1$ points, by its roots, or by its value and all derivative values at a single point. There are many possible representations of a polynomial, all of which may be used as inputs to a computation. Todd (1955) specified a technique for evaluating any polynomial of degree 4 in three multiplications (but five additions) by starting with a different representation of the function. More generally, there is a procedure that will show us how to compute any nth degree polynomial using approximately $n/2$ multiplications. This concept of *preconditioning* is due to Motzkin (1955) and Todd (1955). The development of Theorem 3.1.2 follows

Preconditioning

the exposition in Knuth (1969). (See also Lyusternik et al. [1956], Pan [1966], and Revah [1973] for a survey of results.)

Consider an arbitrary polynomial of degree 6:

$$P_6(x) = \sum_{i=0}^{6} c_i x^i.$$

We may write P_6 as

$$P_6(x) = (x^2 - \alpha_6) P_4(x) + R_6(x),$$

where α_6 is a constant, P_4 is a polynomial of degree 4, and R_6 is of degree at most 1. P_4 may then be written as

$$P_4(x) = (x^2 - \alpha_4) P_2(x) + R_4(x)$$

under similar conditions. Note that the $\{\alpha\}$ are arbitrary. Suppose, then, the $\{\alpha\}$ are chosen somehow so as to make R_6 and R_4 constants (i.e., functions of the $\{c_i\}$ only and not of x). Then $P_6(x)$ may be computed as follows:

(1) Compute x^2 (1 multiplication)
(2) Evaluate $P_2 = b_2 x^2 + b_1 x + b_0$ (2 multiplications)
(3) Compute $P_4 = (x^2 - \alpha_4) P_2(x) + R_4$ (1 multiplication)
(4) Compute $P_6 = (x^2 - \alpha_6) P_4(x) + R_6$ (1 multiplication)

This evaluation then requires only five multiplications.

It is obvious that the factoring may be continued in the same way, yielding an $\lfloor n/2 \rfloor + 2$ multiplication algorithm. (For n odd, we would be computing $x^2, P_1, P_3, \ldots$.) But it remains to be shown that an appropriate set of parameters $\{\alpha_i\}$ can always be obtained. Consider the last step in the proposed scheme:

$$P_n(x) = (x^2 - \alpha_n) P_{n-2}(x) + R_n.$$

Thus,

$$P_n(x) - R_n = (x^2 - \alpha_n) P_{n-2}(x),$$

which means that $\pm \sqrt{\alpha_n}$ are roots of $P_n(x) - R_n$. If we let $\bar{c}_0 = c_0 - R_n$, then

$$\sum_{i=1}^{n} c_i \alpha_n^{i/2} + \overline{c}_0 = 0$$

and

$$\sum_{i=1}^{n} c_i \alpha_n^{i/2}(-1)^i + \overline{c}_0 = 0.$$

Assuming $\alpha_n \neq 0$, these conditions are equivalent to their sum and difference both being 0. That is, the two conditions

(1) $$2\sum_{i=1}^{\lfloor n/2 \rfloor} c_{2i} \alpha_n^i + 2\overline{c}_0 = 0$$

and

(2) $$2\sum_{i=1}^{\lceil n/2 \rceil} c_{2i-1} \alpha_n^{i-\frac{1}{2}} = 0.$$

Condition (1) is no problem, since we can just choose R_n to produce the desired effect. Multiplying (2) by $\sqrt{\alpha_n}$ requires $\sum_{i=1}^{\lceil n/2 \rceil} c_{2i-1} \alpha_n^{i-1} = 0$; i.e., α_n is a root of $\tilde{P}_n(x) \triangleq \sum_{i=1}^{\lceil n/2 \rceil} c_{2i-1} x^{i-1}$. In the same way, α_{n-2} must be a root of $\tilde{P}_{n-2}(x)$. Moreover, $P_n(x) = (x^2 - \alpha_n) P_{n-2}(x) + R_n$ implies that $\tilde{P}_n(x) = (x - \alpha_n) \tilde{P}_{n-2}(x)$; i.e., α_{n-2} is a root of $\tilde{P}_n(x)/(x - \alpha_n)$. Continuing in this manner, we see that $R_4, R_6, \ldots, R_n$ will be constants if and only if $\alpha_4, \alpha_6, \ldots, \alpha_n$ are the roots of $\tilde{P}_n(x)$. If we allow α_i in **C**, then we are done (and this is essentially Motzkin's [1955] result). But if p in **R**$[x]$, we would want $\alpha_i \subset$ **R**. This condition does not necessarily hold for all $P_n(x)$. The following theorem, which we state without proof, allows us to circumvent this problem.

THEOREM 3.1.1 (Eve [1964])

If $P_n(x) = \sum_{i=0}^{n} c_i x^i$ has $n-1$ roots with nonnegative real part, then all the roots of

$$Q_n(x) \triangleq \sum_{i=1}^{\lceil n/2 \rceil} c_{2i-1} x^{i-1}$$

are real.

Preconditioning

We now want to alter P_n so that all its roots will have nonnegative real parts. Consider, then, $\overline{P}_n(x) = P_n(x-r)$, where r is a positive number large enough to guarantee that all roots of $\overline{P}_n(x)$ have positive real parts. (If γ is a root of P_n, then $\gamma + r$ is a root of $\overline{P}_n$.) Now we may use our algorithm on $\overline{P}_n$ with confidence that the α_i are all real. To evaluate $P_n(x)$, we simply compute $\overline{P}_n(x+r)$. With some extra care, then, the following theorem can be obtained (see Problem 3.1).

Theorem 3.1.2 (Knuth [1969])

A polynomial of degree n ($n \geq 3$) in $\mathbf{R}[x]$ can be computed in $\mathbf{R}[x]$, given $\mathbf{R} \cup \{x\}$ in $\lfloor n/2 \rfloor + 2$ multiplications and n additions.

The application of this type of algorithm should be apparent. If a polynomial is to be evaluated at a large number of points, it may be well worthwhile to compute *once* the "preconditioning parameters" $\{\alpha_i, R_i\}$, (i.e., precondition the polynomial). After doing this, a saving of almost $n/2$ multiplications per point can be made over Horner's rule.

In Section 3.2 we will see that the bounds in Theorem 3.1.2 are nearly optimal. However, the required preconditioning process involves finding roots of polynomials. Paterson and Stockmeyer (1973) (motivated by the computation of polynomials whose coefficients and variable are matrices) and Rabin and Winograd (1971) (motivated by round-off considerations) have developed a number of rational preconditioning algorithms in which only rational functions are used in the preconditioning phase. With respect to error analysis, the Rabin and Winograd algorithms are quite promising, even though they warn us that the choice of method for a given polynomial $p(x)$ may depend on the range of values for the variable x. But for low-degree polynomials in common usage, the effort in choosing a method may be well justified. Let us just indicate the method given by Paterson and Stockmeyer.

Theorem 3.1.3

Using rational preconditioning, we can compute any nth degree polynomial with $n/2 + \log n + 0(1)$ multiplications and $3n/2 + 0(1) \pm$ operations.

Proof. Assume $p(x)$ is monic and of degree $n = 2^{k+1} - 1$ for some k (for general n the same bounds can be obtained with a little care). We can decompose $p(x)$ as follows:

$$p(x) = (x^{2^k} + c)p_1(x) + p_2(x)$$

$$= (x^{2^k} + c)(x^{2^k-1} + \sum_{j=0}^{2^k-2} a_j x^j)$$

$$+ (x^{2^k-1} + \sum_{j=0}^{2^k-2} b_j x^j)$$

$$= x^{2^{k+1}-1} + a_{2^k-2} x^{2^{k+1}-2} + \cdots + a_0 x^{2^k}$$

$$+ (c+1)x^{2^k-1} + (ca_{2^k-2} + b_{2^k-2})x^{2^k-2}$$

$$+ \cdots (ca_j + b_j)x^j + \cdots ca_0 + b_0.$$

We need only use rational operations to compute $\{a_j\}$, c, $\{b_j\}$ from the coefficients of $p(x)$. Moreover, we can apply the same idea recursively to compute the monic polynomials $p_1(x)$ and $p_2(x)$, which are both of degree $2^k - 1$. Thus, not counting the cost of the rational preconditioning nor of computing the appropriate powers of x, we can analyze the multiplication cost $f(k + 1)$ and addition cost $g(k + 1)$ to compute $p(x)$ as follows:

$$\left.\begin{array}{r} f(k+1) = 2f(k) + 1 \\ f(1) = 0 \end{array}\right\} f(k+1) = 2^k - 1;$$

$$\left.\begin{array}{r} g(k+1) = 2g(k) + 2 \\ g(1) = 1 \end{array}\right\} g(k+1) = 3 \cdot 2^k - 2.$$

In terms of n, the cost after preconditioning is roughly $n/2 + \log n$ multiplications ($\log n$ for computing the powers of x) and $(3/2)n$ additions. □

Lower Bounds Based on Algebraic Independence 59

Rabin and Winograd (1971) present a number of alternative rational preconditioning methods. One method uses only $n/2 + O(n^{1/2})$ multiplications and $n + O(n^{1/2})$ additions/subtractions. Thus, they show that a total complexity of $3n/2 + o(n)$ can be achieved with rational preconditioning. It remains an open question whether or not rational methods can achieve the $3n/2 + O(1)$ bound of Theorem 3.1.2.

Before leaving this section, we should note that the concept of preconditioning applies to other computations (i.e., other than polynomials in one variable). For example, consider Winograd's (1970a) method for computing, with commutativity, the inner product

$$\sum_{i=1}^{n} a_i b_i = (a_1 + b_2)(a_2 + b_1) + \cdots + (a_{n-1} + b_n)(a_n + b_{n-1})$$
$$- (a_1 a_2 + \cdots + a_{n-1} a_n) - (b_1 b_2 + \cdots + b_{n-1} b_n).$$

Extending this idea to the matrix vector product $A_{n \times n} b_{n \times 1}$ and preconditioning the terms $\sum_{i=1}^{n/2} a_{2i-1,j} a_{2i,j}$ yields a method requiring only $(n^2 + n)/2$ multiplications (but $(3/2)n^2 + (3/2)n$ additions/subtractions) after preconditioning. It follows that the matrix product $A_{n \times n} B_{n \times n}$ can be computed in $n^3/2 + n^2$ multiplications.

3.2 Lower Bounds Based on Algebraic Independence

We shall now show that the upper bound provided by Theorem 3.1.2 is essentially optimal if we are computing any $p(x) \in \mathbf{R}[x]$ whose coefficients are algebraically independent.

DEFINITION 3.2.1

Let H be an extension field of a field F. $u_1, \ldots, u_m \in H$ are *algebraically dependent over* F if there exists a nontrivial multivariate polynomial $f \in F[y_1, \ldots, y_m]$ such that $f(u_1, \ldots, u_m) = 0$. A maximal set of algebraically independent elements in H is called a *transcendence basis* for H over F; the cardinality of such a basis is called the *transcendence degree*.

It is a basic property (van der Waerden [1964, p. 200]) that the transcendence degree is well defined (i.e., in particular, it does not depend on the choice of basis). For a finitely generated extension, this basic property can be stated as a lemma.

LEMMA 3.2.2 (see van der Waerden [1964, p. 200] and Godement [1968, p. 450])

Let $p_1, \ldots, p_m \in H = F(\alpha_1, \ldots, \alpha_t)$. If $m > t$, then $p_1, \ldots, p_m$ are algebraically dependent over F (since $\alpha_1, \ldots, \alpha_t$ is a transcendence basis for H).

Usually, one is concerned with the case that $F = \mathbf{Q}$; in this case, we say that $u_1, \ldots, u_m$ are *algebraically dependent* if there exists a nontrivial $f \in \mathbf{Q}[y_1, \ldots, y_m]$ (or equivalently $f \in \mathbf{Z}[y_1, \ldots, y_m]$) such that $f(u_1, \ldots, u_m) = 0$. Lemma 3.2.2 (with $F = \mathbf{Q}$) is central to the lower bounds of this chapter. We first have to construct certain "canonical programs."

LEMMA 3.2.3 (Motzkin [1955].)

Let P be a program in $F(x)$, given $F \cup \{x\}$, using k * operations. Then we can construct an equivalent program P' which uses (k * operations) and at most $2k$ scalars from F.

Proof. Let $s_i (1 \leq i \leq k)$ denote the result of the ith * operation in P and let s_{k+1} denote the output of P. Then we claim P' can be written in the following canonical form:

$$s_0 \leftarrow x$$
$$\vdots$$
$$s_i \leftarrow [c_i + \sum_{j < i} n_{j,i} s_j] * [d_i + \sum_{j < i} m_{j,i} s_j]$$
$$\vdots$$
$$s_{k+1} \leftarrow c_{k+1} + \sum_{j \leq k} n_{j,k+1} s_j,$$

where $\{c_i\}, \{d_i\} \subseteq F$; $\{n_{j,i}\}, \{m_{j,i}\} \subseteq \mathbf{Z}$.

Scalars introduced between * operations in P are summed together in the program P'. If the first * operation is multiplication, then $s_1 \leftarrow [c_1 + n_{0,1} x] \times [d_1 + m_{0,1} x]$ can be replaced by

Lower Bounds Based on Algebraic Independence

$$s_1 \leftarrow (n_{0,1} m_{0,1})x^2 + (m_{0,1} c_1 + n_{0,1} d_1)x \stackrel{\circ}{=} \bar{n}x^2 + \bar{c}x = \bar{n}x(x + \bar{c})$$

with $\bar{n} \in \mathbf{Z}, \bar{c} \in F$. If the first $*$ operation is division (and assuming we are computing polynomials), then see Problem 3.2(b).

LEMMA 3.2.4 (Belaga [1958])

Let P be a program in $F(x)$, given $F \cup \{x\}$, having $k \pm$ operations. Then we can construct an equivalent program P' which uses at most $k + 1$ scalars from F.

Proof. Let s_i ($1 \leq i \leq k$) denote the result of the ith $\pm$ operation in P and let s_{k+1} denote the output of P. Then we claim P' can be written in the following canonical form:

$$s_0 \leftarrow x$$

$$s_1 \leftarrow c_1 + x^{m_{0,1}}$$

$$\vdots$$

$$s_i \leftarrow c_i + \prod_{j<i} s_j^{m_{j,i}}$$

$$\vdots$$

$$s_{k+1} \leftarrow c_{k+1} \times \prod_{j \leq k} s_j^{m_{j,k+1}}$$

with $\{c_i\} \subseteq F$ and $\{m_{j,i}\} \subseteq \mathbf{Z}$. As in Lemma 3.2.3, it is easy to see how we could combine scalars so that the ith $\pm$ step would be $s_i' \leftarrow c_i' \prod s_j^{n_{j,i}} + d_i' \prod s_j^{m_{j,i}}$. The goal, however, is to introduce only one new scalar rather than two for each $\pm$ operation. Assume we have constructed a program P_{i-1} in which $s_1, \ldots, s_{i-1}$ are of the desired form ($s_l \leftarrow c_l + \prod_{j<l} s_j^{m_{j,l}}$). To construct P_i we will replace $s_i' \leftarrow c_i' \prod s_j^{n_{j,i}} + d_i' \prod s_j^{m_{j,i}}$ by $s_i \leftarrow c_i'/d_i' + \prod s_j^{m_{j,i} \cdot n_{j,i}} \stackrel{\circ}{=} c_i + \prod s_j^{m_{j,i}}$ and then every subsequent occurrence of s_i' in P_{i-1} is replaced by $d_i' \prod s_j^{n_{j,i}} s_i$. □

THEOREM 3.2.5 (Motzkin [1955] and Belaga [1958])

Let P be a program that computes a polynomial $p(x)$ of degree n. If P uses $k < \lfloor (n+1)/2 \rfloor$ $*$ operations or if P uses $k < n$ $\pm$ operations, then the coefficients of $p(x)$ are algebraically dependent.

Lower Bounds and Concepts Related to Algebraic Independence

Proof. Suppose P has k $*$ operations. Then by Lemma 3.2.3 there exists a program P' computing $p(x)$ which uses only $2k$ scalars (call them $\alpha_1, \ldots, \alpha_{2k}$ and think of them as indeterminates). Every statement s_i (and in particular $s_{k+1} = p$) of this program can be expressed as $\sum_j p_j^i(\alpha_1, \ldots, \alpha_{2k}) x^j / \sum_j q_j^i(\alpha_1, \ldots, \alpha_{2k}) x^j$ with each p_j^i and each q_j^i in $Z[\alpha_1, \ldots, \alpha_{2k}] \subseteq Q(\alpha_1, \ldots, \alpha_{2k})$.

To see the basic idea, let us first assume that there were no divisions in P. Then

$$p(x) = \sum_{j=0}^{n} c_j x^j = \sum_{j=0}^{n} p_j^{k+1}(\alpha_1, \ldots, \alpha_{2k}) x^j$$

for some appropriate choice of $\alpha_1, \ldots, \alpha_{2k}$. Using Lemma 3.2.2, we see that either $2k \geq n+1$ or the coefficients, $p_0^{k+1}(\alpha_1, \ldots, \alpha_{2k}), \ldots, p_n^{k+1}(\alpha_1, \ldots, \alpha_{2k})$, are algebraically dependent.

Allowing division, we have

$$p(x) = \sum_{j=0}^{n} c_j x^j = \frac{\sum_{j=0}^{n_1} p_j^{k+1}(\alpha_1, \ldots, \alpha_{2k}) x^j}{\sum_{j=0}^{n_2} q_j^{k+1}(\alpha_1, \ldots, \alpha_{2k}) x^j}$$

for some appropriate choice of $\alpha_1, \ldots, \alpha_{2k}$. Note, however, that not every choice of $\vec{\alpha}$ will make $\sum p_j^{k+1}(\vec{\alpha}) x^j / \sum q_j^{k+1}(\vec{\alpha}) x^j$ a polynomial, and we cannot simply express this rational function over $Q(\vec{\alpha})$ as $\sum_{j=0}^{n} r_j(\vec{\alpha}) x^j$ for some $r_j(\vec{\alpha})$ in $Q(\vec{\alpha})$. We can assume that $\sum p_j^{k+1}(\vec{\alpha}) x^j / \sum q_j^{k+1}(\vec{\alpha}) x^j$ is reduced as a rational function over $Q(\vec{\alpha})$ and that $q_0^{k+1}(\vec{\alpha}) \neq 0$ (else $p(x)$ would be undefined for $x = 0$). Then $1 / \sum_j q_j^{k+1}(\vec{\alpha}) x^j$ can be expanded in a power series $\sum_{j=0}^{\infty} r_j(\vec{\alpha}) x^j$ with the $r_j(\vec{\alpha}) \in Q(\alpha_1, \ldots, \alpha_{2k})$. Thus,

Lower Bounds Based on Algebraic Independence

$$p(x) = [\sum_j p_j^{k+1}(\vec{\alpha})x^j] \times [\sum_j r_j(\vec{\alpha})x^j]$$

$$= \sum_{j=0}^{\infty} \tilde{r}_j(\vec{\alpha})x^j$$

$$= \sum_{j=0}^{n} \tilde{r}_j(\vec{\alpha})x^j \quad \text{(By taking the previous equality mod } x^{n+1}\text{).}$$

Lemma 3.2.2 again applies, completing the argument.

The result for ± operations follows in exactly the same way.

□

We can restate Theorem 3.2.5 as follows: "Most" (the "exceptions" are contained in a countable union of algebraic subsets of $\mathbf{R}^{n+1}$ space and hence have measure zero) nth degree polynomials require $\geq \lceil (n+1)/2 \rceil$ * operations and $\geq n$ ± operations. Of course these exceptions may be just the only polynomials we care to compute; for, if any of the coefficients is rational (or even just algebraic), we cannot apply Theorem 3.2.5 to obtain a lower bound.

We close this section with a short discussion of how ÷ is accommodated in Strassen (1974). This approach is not necessary here, but is needed for a complete proof of Theorems 3.3.6 and 3.4.4. For a program

$$P = \begin{matrix} s_1 \\ \vdots \\ s_t \end{matrix}$$

with u ± operations and v multiplications, we know that every step s_i (and, in particular, s_t) can be represented as $\sum_{j=0}^{r} p_j^i(\alpha_1, \ldots, \alpha_m)x^j$, where the number of parameters m can be bounded by min $[u+1, 2v]$. Indeed, the polynomials $p_j^i \in \mathbf{Z}[y_1, \ldots, y_m]$ can be defined by induction on i (= the number of steps). Moreover, with additional care, this same type of induction can be extended to computations in the power series ring $F[[x]]$ allowing ÷ by units in $F[[x]]$; i.e.,

$$f(x) = \sum_{i=0}^{\infty} b_i x^i$$

can be a divisor as long as $b_0 \neq 0$.

Now suppose we are computing $p(x) = \sum_{j=0}^{n} a_j x^j$ by a program P in $F(x)$; i.e., allowing division. In P, we interpret each step s_i ($1 \leq i \leq t$) as an element of $F(x)$; we could also interpret (and compute) each s_i as an element of the power series ring $G[[\tilde{x}]]$, where $G = F(\Theta)$, $\tilde{x} = x - \Theta$, and Θ is a new indeterminate. We can compute $p(x) \stackrel{\circ}{=} \tilde{p}(\tilde{x})$ by a program P' in $G[[\tilde{x}]]$, given $G \cup \{\tilde{x}\}$ with

$$P' = \begin{cases} x = s_0 \leftarrow \tilde{x} + \Theta \\ P \end{cases}$$

We have

$$\tilde{p}(\tilde{x}) = \sum_{j=0}^{n} \tilde{a}_j \tilde{x}^j = \sum_{j=0}^{n} a_j x^j = p(x)$$

with $\tilde{a}_j = \sum_{k=j}^{n} \binom{k}{k-j} a_k \Theta^{k-j}$. As before, the coefficients $\tilde{a}_0, \ldots, \tilde{a}_n$ will be algebraically dependent if P' uses fewer than $n \pm$ operations or $(n/2) *$ operations. That is, $f(\tilde{a}_0, \ldots, \tilde{a}_n) = 0$ for some nontrivial f in $Z[y_0, \ldots, y_n]$. We can view $f(\tilde{a}_0, \ldots, \tilde{a}_n)$ as a polynomial $q(\Theta)$ with coefficients in $Z[a_0, \ldots, a_n]$. Since $q(\Theta)$ is identically 0, it follows that the constant term of $q(\Theta)$, $f(a_0, \ldots, a_n)$, is 0. That is, the coefficients of $p(x)$, which is computed by P in $F(x)$, are algebraically dependent and are zeros of the same f in $Z[y_0, \ldots, y_n]$.

3.3 Polynomials with Rational Coefficients

It can be argued that only polynomials in $Q[x]$ are of any computational concern, and we have already noted that the lower bounds of Theorem 3.2.5 are not applicable in this case. Indeed, when computing over $Q[x]$, a program with k scalar $*$ operations can be replaced by a program with k integer multiplications and (possibly) one final scalar (= rational) multiplication. The integer multiplications can then be simulated by a number (not bounded by the degree of the polynomial being computed) of $\pm$ operations. The next theorem shows that relatively few nonscalar $*$ operations are required.

Polynomials with Rational Coefficients

THEOREM 3.3.1 (Paterson and Stockmeyer [1973])

Let $p(x)$ be any nth degree polynomial in an arbitrary domain $F[x]$. Then $p(x)$ can be computed in $F[x]$, using only $\sqrt{2n} + 0(\log n)$ nonscalar multiplications.

Proof. It is easy to see how to "factor" $p(x)$ to obtain a $2\sqrt{n}$ method. For example,

$$\sum_{i=0}^{8} c_i x^i = ((c_8 x^2 + c_7 x + c_6) x^3 + (c_5 x^2 + c_4 x + c_3))x^3$$

$$+ (c_2 x^2 + c_1 x + c_0).$$

The better bound of $\sqrt{2n} + 0(\log n)$ is achieved by an appropriate factoring of the method indicated in Theorem 3.1.3, rather than Horner's rule. □

Paterson and Stockmeyer also show that $\sqrt{n}$ nonscalar * operations are required. Again, we will first establish the result under the restriction that ÷ is not allowed.

LEMMA 3.3.2

Let P be a program in $F[x]$ using k nonscalar multiplications. Then there is an equivalent program P' which uses only $k^2 + 1$ scalars.

Proof. P' will be in the following form (see also Theorem 2.4.7):

$$s_{-1} \leftarrow 1$$

$$s_0 \leftarrow x$$

$$\vdots$$

$$s_i \leftarrow [\sum_{-1 \leq j < i} \alpha'_{j,i} s_j] \times [\sum_{-1 \leq j < i} \alpha''_{j,i} s_j]$$

$$\vdots$$

$$s_{k+1} \leftarrow \sum_{j \leq k} \alpha'_{j,k+1} s_j$$

where $\alpha'_{j,i}, \alpha''_{j,i} \subseteq F$.

Actually, this canonical form uses

$$\sum_{i=1}^{k} (2i + 2) = k^2 + 4k + 2$$

parameters $\{\alpha'_{j,i}, \alpha''_{j,i}\}$. With a little more care the number of parameters can be reduced to $k^2 + 1$ (e.g., we can set $\alpha'_{j,-1} = \alpha''_{j,-1} = 0$ for $1 \leq j \leq k$).

□

It is well worth noting the similarity between Lemma 3.3.2 and Lemma 3.2.3. Here the integers $\{n_{j,i}\}$ and $\{m_{j,i}\}$ in Lemma 3.2.3 have been replaced by the scalar parameters $\{\alpha'_{j,i}\}$ and $\{\alpha''_{j,i}\}$. We now have $\sim k^2$ parameters rather than $2k$. On the other hand, we now have only *one* fixed canonical program rather than a countable class of canonical programs (one program for each choice of the integers $\{n_{j,i}\}, \{m_{j,i}\}$).

We are interested in the computation of $p \in F[x]$ in $G[x]$, given $G \cup \{x\}$, where $F \subseteq G$ and x is indeterminate with respect to G (e.g., $F = \mathbf{R}$, $G = \mathbf{C}$).

THEOREM 3.3.3 (Paterson and Stockmeyer [1973]

There exists a nontrivial f in $\mathbf{Z}[y_0, \ldots, y_n]$ such that any polynomial $\sum_{i=0}^{n} c_i x^i$ in $F[x]$ which can be computed in $G[x]$, given $G \cup \{x\}$, in less than $\sqrt{n}$ nonscalar multiplications satisfies $f(c_0, \ldots c_n) = 0$.

Proof. Let P be *the* canonical k nonscalar × step program. Each s_i (and in particular s_{k+1}) can be expressed as $\sum_j p_j(\alpha)x^j$. Suppose $k < \sqrt{n}$ (i.e., $k^2 + 1 < n + 1$). Then $\{p_0(\alpha), \ldots, p_n(\alpha)\}$ are algebraically dependent and hence there exists a nontrivial $f \in \mathbf{Z}[y_0, \ldots, y_n] : f(p_0(\alpha), \ldots, p_n(\alpha)) = 0$. Therefore, every nth degree polynomial $\sum_{i=0}^{n} c_i x^i$ computable in $k < \sqrt{n}$ nonscalar multiplications is such that $f(c_0, \ldots, c_n) = 0$.

□

COROLLARY 3.3.4

"Most" $p(x) \in \mathbf{Q}[x]$ of degree n require $\geq \sqrt{n}$ nonscalar multiplications (computing p in $\mathbf{C}[x]$, given $\mathbf{C} \cup \{x\}$).

Polynomials with Rational Coefficients

Proof. In order to give "most" a precise meaning, we can identify $\{p \mid p \in Q[x], \deg p = n\}$ with O^{n+1} via

$$\sum_{i=0}^{n} c_i x^i \overset{1\text{-}1}{\leftrightarrow} \langle c_0, \ldots, c_n \rangle.$$

Then the set $S = \{\langle c_0, \ldots, c_n \rangle \mid \sum_{i=0}^{n} c_i x^i$ computable in fewer than $\sqrt{n}$ nonscalar multiplications$\}$ is nowhere dense in $\mathbf{R}^{n+1}$. For let f be as in Lemma 3.3.2 and suppose S were dense in some open set $U \subseteq \mathbf{R}^{n+1}$, then $f(S) = 0 \Rightarrow f(U) = 0$ (by continuity) $\Rightarrow f \equiv 0$ (since U is open in $\mathbf{R}^{n+1}$).

□

As in Section 3.2, a few additional remarks will allow us to extend the results to show that $\sqrt{n}$ nonscalar $\times, \div$ operations are required. If $\div$ is allowed, then there will be 2^k canonical k step programs, one program for each of the 2^k ways to choose the k * operations. Polynomials computed by the lth canonical program can again be expressed as $\sum_{j=0}^{n} r_j^l(\alpha_1, \ldots, \alpha_{k^2+1}) x^j$ for $1 \leq l \leq 2^k$. If $k < \sqrt{n}$, then $\{r_0^l(\vec{\alpha}), \ldots, r_n^l(\vec{\alpha})\}$ are algebraically dependent and there exist nontrivial $f_l \in Z[y_0, \ldots, y_n] : f_l(r_0^l(\vec{\alpha}), \ldots, r_n^l(\vec{\alpha})) = 0$. Now consider $f \overset{\circ}{=} \prod_{l=1}^{2^k} f_l$. By the definition, the $n + 1$ coefficients of any nth degree polynomial computable in fewer than $\sqrt{n}$ nonscalar * operations constitute a zero of the nontrivial polynomial f. Theorem 3.3.3 and Corollary 3.3.4 thereby generalize to computations in $F(x)$.

Thus, the distinction observed by Paterson and Stockmeyer is the difference between a finite set and a countable set of canonical programs. The next result helps to emphasize this distinction.

THEOREM 3.3.5

For any n and t, there exist nth degree polynomials p in $\mathbf{Q}[x]$ such that every program P which computes p in $\mathbf{C}(x)$, given $\mathbf{C} \cup \{x\}$ either (1) uses more than t total operations or (2) uses at least $\lfloor n/2 \rfloor + 1$ * operations and n ± operations.

The proof follows as a corollary to the ideas leading to Corollary 3.3.4; for, given a bound on the total number of operations, there are only a finite number of canonical $k *$ step programs (and canonical $k \pm$ step programs). We see, therefore, that the exorbitant cost associated with the simulation of (integer) scalar multiplication by addition is necessary if we want to compute all $p \in \mathbf{Q}[x]$ in $\leq \lfloor n/2 \rfloor *$ operations.

In a sense, the preceding remarks make the $\sqrt{2n} \times$ upper bounds (for all $p \in \mathbf{Q}[x]$) seem a little more surprising. In contrast, we cannot prove an analogous upper bound for $\pm$ operations; i.e., it may be the case that "most" nth degree $p \in \mathbf{Q}[x]$ require $n \pm$ operations (no matter how many $*$ operations are used). The best we can do at present is to prove an analogue to Lemma 3.3.2 (and its consequences, the lower bounds of Theorem 3.3.3 and Corollary 3.3.4).

THEOREM 3.3.6 (Borodin and Cook [1974])

Let P be a program in $F(x)$ using $k \pm$ operations and assume P computes nth degree polynomial $p(x) \in F[x]$. There is a bound $b(n,k)$ and rational functions $r_j^l(\alpha_1, \ldots, \alpha_t)$ in $\mathbf{Q}(\alpha_1, \ldots, \alpha_t)$, $0 \leq j \leq n$, $1 \leq l \leq b(n,k)$, $t = (k+2)^2$ such that p can be represented as $\sum_{j=0}^{n} r_j^l(\alpha_1, \ldots, \alpha_t) x^j$ for some l and some choice of $\alpha_1, \ldots, \alpha_t$ in F.

Sketch of proof. Recall the canonical program in Lemma 3.2.4. We can accommodate $\div$ if we allow (possibly negative) exponents $m_{j,i}$ from $\mathbf{Z}$ (rather than just $\mathbf{N}$). On the other hand, we will have to view the computation as taking place in some $G[[\widetilde{x}]]$, $G = F(\Theta)$, and $\widetilde{x} = x - \Theta$ so that all s_j^{-u} $(u > 0) = (s_j^{-1})^u$ can be expanded in a power series (see the concluding remarks of Section 3.2; specifically, if $\widetilde{p}(\widetilde{x}) = p(x)$ and the coefficients of $\widetilde{p}$ are algebraically dependent, then the coefficients of p will be algebraically dependent). To simplify the discussion, we will assume that all exponents $\{m_{j,i}\}$ are nonnegative, $\Theta = 0$, and the computation is taking place in $F[x]$.

Unlike the situation for nonscalar $*$ operations, we cannot just let the $\{m_{j,i}\}$ be parameters and have every step be represented as $\sum p_j(m) x^j$. Let's concern ourselves only with the computation of nth degree polynomials (and think of n as being fixed). Suppose a program P in $F[x]$ computes p. Then P correctly computes p in $F[x] \bmod x^{n+1}$; i.e., all terms of degree $\geq n+1$ are dropped throughout the computation. The example $s \leftarrow (x + \alpha)^u$ illustrates the approach to be taken. The goal is to represent $s \bmod x^{n+1}$ as

Polynomials with Rational Coefficients

$\sum_{j=0}^{n} r_j(\alpha_1, \alpha_2, \ldots, \alpha_r) x^j$ for some finite set of parameters $\alpha_1, \ldots, \alpha_r$ and r_j in $Z[y_1, \ldots, y_r]$. We work by cases:

(i) $u = i$, $0 \leq i \leq n$. In each of these $n + 1$ cases, it is clear that we can represent $s \mod x^{n+1}$ as some $\sum_{j=0}^{n} r_j(\alpha) x^j$

(ii) $u \geq n + 1$:
$$s \mod x^{n+1} = \alpha^u + u\alpha^{u-1}x + \cdots + \binom{u}{n}\alpha^{u-n} x^n$$
$$= \sum_{j=0}^{n} r_j(\alpha_1, \alpha_2, \alpha_3) x^j$$

where $\alpha_1 = \alpha$, $\alpha_2 = \alpha^u$, $\alpha_3 = u$, and
$r_0(\alpha_1, \alpha_2, \alpha_3) = \alpha_2$
$r_1(\alpha_1, \alpha_2, \alpha_3) = \alpha_3 \alpha_2 / \alpha_1$
$\vdots$
$r_n(\alpha_1, \alpha_2, \alpha_3) = \binom{\alpha_3}{n} \dfrac{\alpha_2}{\alpha_1^n}$

The proof is by induction on k, arguing by cases depending on whether or not any $m_{j,i} \leq n$ or $> n$. A $k \pm$ step program introduces $v = ((k+1)(k+2))/2$ exponents, all of which we will treat as parameters. For every exponent there are $n + 2$ cases to consider ($m_{j,i} = 0$, $m_{j,i} = 1, \ldots$), or $(n + 2)^v$ cases in all. Each case will determine a new canonical program. For each of these (finite number of) programs, we will characterize the statements in the desired manner.

Let's just consider the case that all exponents are greater than n (of course we could argue trivially that we are not computing p, but this approach shows that we are not even computing $p \mod (x^{n+1})$ if k is too small).

Induction step. Assume

$$s_i = \sum_{j=0}^{n} p_j^i(\alpha_1, \ldots, \alpha_{t(i)}) x^j \mod (x^{n+1}) \qquad \text{for } 0 \leq i \leq r.$$

Show that

$$s_{r+1} = \sum_{j=0}^{n} p_j^{r+1}(\alpha_1, \ldots, \alpha_{t(r+1)}) x^j \mod (x^{n+1})$$

and that $t(r+1) \leq t(r) + 2(r+1)$. So, by induction, $t = t(k+1) \leq (k+2)^2$.

Introduce new parameters (and rename by $\alpha_{t(r)+1}, \alpha_{t(r)+2}, \ldots, \alpha_{t(r+1)}$ to represent $c_{r+1}, m_{0,r+1}, \ldots, m_{r,r+1}, [p_0^1(\alpha_1, \ldots, \alpha_{t(1)})]^{m_{1,r+1}}, \ldots, [p_0^r(\alpha_1, \ldots, \alpha_{t(r)})]^{m_{r,r+1}}$. We have thus introduced $2(r+1)$ parameters. Now it remains to show that

$$s_{r+1} = \Sigma \, p_j^{r+1}(\alpha_1, \ldots, \alpha_{t(r+1)}) x^j \pmod{x^{n+1}}.$$

Now $\quad s_{r+1} = \prod_{j=0}^{r} s_j^{m_{j,r+1}} + c_{r+1}$, so

look at any

$$s_i^{m_{i,r+1}} = \left[\sum_{j=0}^{n} p_j^i(\alpha_1, \ldots, \alpha_{t(i)}) x^i \right]^{m_{i,r+1}} \mod(x^{n+1}).$$

Claim. The coefficient of x^l ($l \leq n \leq m_{i,r+1}$) is

$$\sum_{m_1 \cdot 1 + m_2 \cdot 2 + \cdots + m_n \cdot n = l} \binom{m_{i,r}+1}{m_1} \binom{m_{i,r+1}-m_1}{m_2} \cdots$$

$$\binom{m_{i,r+1}-m_1-\cdots-m_n}{m_n}$$

$$\left[p_0^i(\alpha_1, \ldots) \right]^{m_{i,r+1}-m_1\cdots m_n} p_1^i(\;)^{m_1} \cdots p_n^i(\;)^{m_n}.$$

and, as in the simple example, the expression can be written as a rational function $g(\alpha_1, \ldots, \alpha_{t(r)}, m_{i,r+1}, [p_0^i]^{m_{i,r+1}})$. So it follows that $\prod s_j^{m_{i,r+1}} + c_{r+1} \mod (x^{n+1})$ can be represented as desired. □

COROLLARY 3.3.7

"Most" $p(x) \in \mathbf{Q}[x]$ of degree n require $\geq \sqrt{n+1} - 2 \pm$ operations (computing p in $\mathbf{C}(x)$ given $\mathbf{C} \cup \{x\}$).

Proof. If p is computed in $k \pm$ operations, then we can view p as being computed by one of a finite number, $b(n,k)$, of programs.

Specific Polynomials That are Difficult to Compute 71

The nth degree polynomials computed by the lth of these programs can be represented as $\sum_{j=0}^{n} r_j^l(\alpha_1, \ldots, \alpha_t)x^j$ with $t = (k+2)^2$. If $t < n+1$, then there exists f_l in $Z[y_0, \ldots, y_n]$ such that $f_l(r_0^l, \ldots, r_n^l) = 0$. Then, as in the proof of (and the remarks following) Corollary 3.3.4, the $n+1$ coefficients of any nth degree polynomial computable in k ($k < \sqrt{n+1} - 2$) $\pm$ operations would constitute a zero of $f \triangleq \prod_{l=1}^{b(n,k)} f_l$.

□

We don't see how $*$ operations can in general reduce the $\pm$ complexity (i.e., so that all $p \in Q[x]$ are computable in $O(\sqrt{n}) \pm$ operations). It is tempting to conjecture the existence of a function $\psi(k,n)$ satisfying the following property: If p is an nth degree polynomial computable in $k \pm$ operations, then p is computable in $k \pm$ operations and $\leq \psi(k,n) *$ operations.

3.4 Specific Polynomials That Are Difficult to Compute

In Sections 3.2 and 3.3 we established the existence of computationally difficult polynomials. We would like to be able to explicitly specify concrete examples (say in $Z[x]$). The crux of the proofs based on algebraic independence is to show that every step s_i can be represented as $\sum_{j=0}^{r} p_j^i(\alpha_1, \ldots, \alpha_{t(i)})x^j$ with $t(i)$ not growing too fast. A more careful analysis yields bounds on the degree and coefficient size of the polynomials p_j^i in $Z[y_1, \ldots, y_{t(i)}]$. If $p(x) = \sum_{j=0}^{n} p_j(\alpha_1, \ldots, \alpha_t)x^j$ and $t < n+1$, then the coefficients of p are algebraically dependent; furthermore, a bound on the degree and coefficient size of the p_j determines a bound on the degree and coefficient size of some $h(z_0, \ldots, z_n)$ whose zeros include $\langle p_0(\vec{\alpha}), \ldots, p_n(\vec{\alpha}) \rangle$ (i.e., h exhibits the algebraic dependence of the $\{p_j(\vec{\alpha})\}$). The results in this section are due to Strassen (1974).

DEFINITION 3.4.1

Let $p(y_1, \ldots, y_q) = \Sigma\, c_{n_1 \ldots n_q} y_1^{n_1} \cdots y_q^{n_q}$

(a) Height $p \stackrel{\circ}{=} \max(|c_{n_1 \ldots n_q}|)$.
(b) Weight $p \stackrel{\circ}{=} \Sigma |c_{n_1 \ldots n_q}|$.

LEMMA 3.4.2

Let
$$P = \begin{matrix} s_1 \\ \vdots \\ s_t \end{matrix}$$

be a program (which we think of as computing an nth degree polynomial) in $F[[x]]$, given $F \cup \{x\}$; and let $s_i = \sum_{j \geq 0}^{n} c_{ji} x^i$ (mod x^{n+1}). Suppose there are $u \pm$, $v_1 \times$ and $v_2 \div$ (by units) operations in P. Let $v = v_1 + v_2$ and $m = \max(u, 2v)$. Then there exist polynomials p_j^i in $Z[y_1, \ldots, y_m]$, $\max_{1 \leq j \leq t} \deg p_j^i \leq d \stackrel{\circ}{=} (u+1)2^{v_1+1} n^{v_2}$, $\sum_{1 \leq j \leq n}$ weight $p_j^i \leq 2^d$ and $\lambda_1, \ldots, \lambda_t, \alpha_1, \ldots, \alpha_m$ in $F: c_{ji} = \lambda_i p_j^i(\alpha_1, \ldots, \alpha_m)$ $1 \leq j \leq n$, $1 \leq i \leq t$ (if $m = 2v$, then $\lambda_i = 1$).

Sketch of proof. Arguing separately for the cases $m = u$ and $m = 2v$, the lemma is proved by induction on t. The existence of the p_j^i has already been suggested; the bounds on degree and weight are new here. Let us just look at the case $m = 2v$ and assume no $\div$ (i.e., $v = v_1$).

Initially, we have elements $c = c + (0 \cdot x + \cdots)$ in F and $x = 1 \cdot x + 0 \cdot x^2 + \cdots$. If $s_t = s_k \pm s_l$, then $p_j^t = p_j^k \pm p_j^l$ for $j \geq 1$. If $s_t = s_k \times s_l$, then p_j^t is defined by equating coefficients of x^j in

$$\sum_{j \geq 1} p_j^t(\alpha_1, \ldots, \alpha_{2v}) x^j + \alpha_{2v-1} \alpha_{2v}$$

$$= [\alpha_{2v-1} + \sum_{j \geq 1} p_j^k(\alpha_1, \ldots, \alpha_{2v-2}) x^j] \times$$

$$[\alpha_{2v} + \sum_{j \geq 1} p_j^l(\alpha_1, \ldots, \alpha_{2v-2}) x^j].$$

Specific Polynomials That are Difficult to Compute

For ±,

$$\deg p_j^t \leq \max(\deg p_j^k, \deg p_j^l) \leq (u+1)2^{v_1+1}$$

$$\Sigma \text{ weight } p_j^t \leq \Sigma \text{ weight } p_j^k + \Sigma \text{ weight } p_j^l$$

$$\leq 2 \cdot 2^{(u+1)2^{v_1+1}}$$

$$\leq 2^{(u+2)2^{v_1+1}}$$

For ×,

$$\deg p_j^t \leq \deg p_j^k + \deg p_j^l$$

$$\leq 2(u+1)2^{v_1+1} = (u+1)2^{v_1+2}$$

$$\Sigma \text{ weight } p_j^t \leq \Sigma \text{ weight } p_j^k \cdot \Sigma \text{ weight } p_j^l$$

$$\leq [2^{(u+1)2^{v_1+1}}]^2$$

$$= 2^{(u+2)2^{v_1+2}}$$

If we let $s = u + v =$ total operations, and assume $n \geq 5, u \geq 1$, then we can express the bounds as

$$\deg p_j^t \leq (u+1)2^{v_1+1} n^{v_2}$$

$$\leq 2(u+1)n^v \leq n^s,$$

$$\text{weight } p_j^t \leq \sum_{1 \leq j \leq n} \text{weight } p_j^t \leq 2^{n^s}.$$

LEMMA 3.4.3 (note the relationship to Lemma 3.2.2)

Let $p_1, \ldots, p_q$ be in $\mathbf{Z}[\alpha_1, \ldots, \alpha_m]$ with $\deg p_j \leq d$, weight $p_j \leq w$ for $1 \leq j \leq q$. Assume $m \geq 1, d \geq 2, w \geq 4$ and $q \geq 5$. Suppose g is large enough to satisfy $g^{q-m-2} > d^m q^q \log w$. (Note that we implicitly have $q > m + 2$.) Then there is an integer form (i.e., homogeneous polynomial with integer coefficients) h of degree g and height ≤ 3 such that $h(p_1, \ldots, p_q) = 0$.

Sketch of proof. The idea is to express the problem in terms of finding a nontrivial zero for $M < N$ linear forms in $Z[Y_1, \ldots, Y_N]$, each of weight $\leq G$. We are guaranteed an integer solution (for $Y_1, \ldots, Y_N$) with $|Y_i| \leq G^{M/N-M} + 2$. (A proof can be found in Schneider [1957, p. 140].)

In this case, we want to find integers $y_{i_1, \ldots, i_q}$ such that

$$\sum_{i_1 + \cdots + i_q = g} y_{i_1 \ldots i_q} p_1^{i_1} \cdots p_q^{i_q} = 0.$$

Thinking of $Y_{i_1 \ldots i_q}$ as indeterminates, we have

$$Q'(\vec{\alpha}) = \Sigma \, Y_{i_1 \ldots i_q} p_1^{i_1}(\vec{\alpha}) \cdots p_q^{i_q}(\vec{\alpha}),$$

where the coefficients of Q' are linear forms in $Z[\vec{Y}]$ of weight $\leq w^g \binom{g+q-1}{q-1}$ and Q' has degree $\leq gd$. There are at most $M = \binom{gd+m}{m}$ such coefficients (linear forms) in the $N = \binom{g+q-1}{q-1}$ variables $Y_{i_1 \ldots i_q}$. The $y_{i_1 \ldots i_q}$ are a common zero of these M linear forms, and the assumptions on g, d, m, q, w are sufficient to permit $|y_{i_1 \ldots i_q}| \leq 3$. □

THEOREM 3.4.4

Let P be a program in $F[[x]]$ computing $\sum_{j=0}^{n} a_j x^j$ using $u \pm$ operations and $v *$ operations; let $m = \min(u, 2v)$ and $s = u + v$, and assume that $q > m + 2$. Then there is a nontrivial integer form h of degree g and height ≤ 3 such that $h(a_{i_1}, \ldots, a_{i_q}) = 0$ whenever g satisfies $g^{q-m-2} > (n^s)^{m+1} q^q$.

The proof follows directly from Lemmas 3.4.2 and 3.4.3. By the remarks at the end of Section 3.2, we can draw the same conclusion for computations in $F(x)$. But not all algebraically dependent coefficients satisfy "simple" algebraic relations (i.e., polynomials of low degree and height).

LEMMA 3.4.5

Let $a_1, \ldots, a_q$ in Z satisfy the following conditions: $|a_1| > 4$, $|a_i| > |q \cdot a_{i-1}|^g$. Then there is no integer form of degree g and height ≤ 3 such that $h(a_1, \ldots, a_q) = 0$.

COROLLARY 3.4.6

Let $p(x) = \sum_{i=0}^{n} a_i x^i = \sum_{i=0}^{n} 2^{2^{in^3}} \cdot x^i$. Then any program P in $C(x)$, given $C \cup \{x\}$ which computes $p(x)$ must have $m = \min(u, 2v) \geq n - 4$ or $s > n^2/\log n$. That is, the program either uses close to $n \pm$ operations *and* $n/2 *$ operations or it uses $n^2/\log n$ total operations (a trade-off one is not likely to take).

Proof. We have $a_i = 2^{2^{in^3}}$. Let $g = 2^{n^3 - n^2}$ and verify that $|a_1| > 4$ and $|a_i| > |na_{i-1}|^g$. Then, by Lemma 3.4.5, there can be no form h of degree g and height 3 such that $h(a_1, \ldots, a_n) = 0$. Theorem 3.4.4 then dictates that m and s must satisfy $g^{n-m-2} \leq n^{s(n+1)} n^n$, which implies (taking logarithms) $m \geq n - 4$ or $s > n^2/\log n$. □

We know from Section 3.3 that $p(x)$ can be computed in $O(\sqrt{n}) *$ operations; now we know that this reduction in $*$ operations will be quite costly.

PROBLEMS

3.1 The discussion preceding Theorem 3.1.2 yields a method for preconditioned polynomial evaluation using $\lfloor n/2 \rfloor + 2$ multiplications and $n + 1 \pm$ operations. Show how to achieve the bound stated in Theorem 3.1.2.

3.2 For preconditioned polynomial evaluation in $C(x)$, the bound for the number of $*$ operations is now known exactly.
  (a) (Motzkin [1955].) Show how to achieve an upper bound of $\lceil (n+2)/ \rceil$ rather than $\lceil (n+3)/2 \rceil$ as suggested in the discussion preceding Theorem 3.1.1 By Theorem 3.2.5, this bound is optimal for n even.
  (b) (Revah [1973].) With regard to Lemma 3.2.3, show that if the first $*$ operation is division, and every subsequent $*$ operation introduces two scalars (algebraically independent with respect to scalars previously introduced), then the program is not computing a polynomial.
  (c) (Knuth [1969].) For $n = 3, 5,$ or 7, show that the bound $\lceil (n+2)/2 \rceil$ is optimal.
  (d) (Revah [1973].) For n odd and $n \geq 9$, show that the bound $\lceil (n+1)/2 \rceil$ is optimal.

3.3 Let $p(x) = \sum_{i=0}^{n} c_i x^i$ with $c_0, \ldots, c_n$ in **R** algebraically independent. Suppose we have to compute $p(x_1), \ldots, p(x_m)$ iteratively (i.e., x_{j+1} is not known until $p(x_j)$ has been computed). Show that at least $(3/2)nm$ arithmetic operations are required.

3.4 Show that there exist matrices B in $\mathbf{R}_{n \times m}$ such that $\vec{B}\vec{x}_{m \times 1}$ requires $(3/2)nm$ arithmetic operations.

3.5 (Winograd [1969].) Suppose we want to compute $\sum_{i=1}^{n} x_i y_i$ in $F(\vec{x},\vec{y})$ given $F(\vec{x}) \cup F(\vec{y})$. Show that $2n - 1$ total operations are required even though only $\lceil n/2 \rceil$ multiplications are sufficient.

3.6 Complete the proof of Theorem 3.3.1; i.e., provide a factorization of any nth polynomial so that it can be computed in $\sqrt{2n} + O(\log n)$ nonscalar multiplications.

3.7 (Paterson and Stockmeyer [1973].) Use a counting argument and the homomorphism on programs induced by $Z \to Z_2$ to show that there are nth degree polynomials with $\{0,1\}$ coefficients whose computation in $Z[x]$ requires $\sqrt{n} - 1$ nonscalar multiplications.

3.8 Complete the proof of Lemma 3.4.2.

CONJECTURES—OPEN PROBLEMS

1. Can the lower bounds for rational preconditioning be improved? In particular, must any such method require $(n/2) + c \log n$ multiplications (not counting preconditioning of course)?

2. Exhibit a specific nth degree polynomial with coefficient in $\{0,1\}$ which is difficult to compute.

Chapter 4

THE FAST FOURIER TRANSFORM, RELATED CONCEPTS, AND APPLICATIONS

The discrete Fourier transform can be viewed as a transformation from the representation of a polynomial by its coefficients to the representation by its values at a suitable number of special points. These points are the roots of unity and the fast Fourier transform (see Cooley and Tukey [1965], Cooley, Lewis, and Welch [1967], Gentleman and Sande [1966], and Knuth [1969, p. 441]) is an "efficient" method of performing this transformation in both the forward and inverse directions.

In this chapter we present and apply the FFT (fast Fourier transform) to produce some surprising upper bounds. As a standard example, consider the problem of multiplying two nth degree polynomials to obtain the $2n + 1$ coefficients of the product polynomial. How many arithmetic operations are required? The classical algorithm is $O(n^2)$, but the FFT technique yields an $O(n \log n)$ algorithm. This technique has also been applied by Schönhage and Strassen (1971) to produce an $O(n \log n \log \log n)$ "bit operations" method for multiplying two n bit numbers, as compared to the $O(n^2)$ classical method.

Modular arithmetic techniques are completely analogous to the use of the discrete Fourier transform. We will begin with a brief discussion of modular arithmetic and some strategies for developing fast algorithms.

4.1 Modular Arithmetic and Strategies for Fast Algorithms

Recall the important Chinese remainder theorem.

THEOREM 4.1.1 (This is not the most general statement possible.)
Let $p_1, p_2, \ldots, p_n$ be relatively prime. Then for any $r_1, \ldots, r_n$ ($r_i < p_i$), there is a unique nonnegative integer u satisfying $u < \prod_{i=1}^{n} p_i$ and $r_i = u \bmod p_i$ for $1 \leq i \leq n$.

The Chinese remainder theorem and the following rules of arithmetic provide the basis for modular techniques.

FACT 4.1.2

   (a) $(x + y) \bmod a = (x \bmod a + y \bmod a) \bmod a$.
   (b) $(xy) \bmod a = [(x \bmod a)(y \bmod a)] \bmod a$.
   (c) For any p, Z_p = integers mod p under the operations $+$ and $\times$ is a ring. In particular, $-x$ is the unique $y < p$ such that $x + y = 0 \bmod p$.
   (d) If p is a prime, then Z_p is a field. In particular, x^{-1} is well defined for every $x \neq 0 \bmod p$; it is, of course, the unique $y < p$ such that $xy = 1 \bmod p$.

Modular techniques have been used in a wide variety of applications (see, for example, the seminal paper by Takahasi and Ishibashi [1961], the extensive development of modular algorithms by Collins [1971] in the SAC-1 system, the discussion in Knuth [1969], and the survey paper by Horowitz [1971]). These techniques are most beneficial in problems in which intermediate calculations may give rise to numbers much larger than the inputs and outputs.

Suppose we wish to multiply two matrices whose elements are single precision integers. The product $C = A \times B$ has entries $c_{ij} \leq na^2$, where $a = \max_{i,j} (|a_{ij}|, |b_{ij}|)$. Hence, c_{ij} is at most double or triple precision, assuming n is not too large with respect to a. It is possible that a method which is fast in terms of arithmetic operations may give rise to large intermediate numbers (this turns out *not* to be the case with Strassen's (1969) method), and hence the cost of multiple precision arithmetic (numbers of size a^n are n precision) can swamp any gains made in reducing the number of operations. Since the final answers for our matrix product are at worst triple precision, the entire computation can be carried out mod p for some large enough triple precision prime $p > na^2$. Alternatively, the computation can be carried out mod p_k, $1 \leq k \leq 3$ for three single precision primes p_k such that $p_1 p_2 p_3 > na^2$. If $c_{ij}^{(k)}$ is the i,jth entry of C (when the calculation is done mod p_k), then c_{ij} is the unique $u < p_1 p_2 p_3$ such that $c_{ij}^{(k)} = u \bmod p_k$. Hence, although we have tripled the number of operations, *each operation is single precision.*

What do multiple precision arithmetic and modular representations have to do with polynomials? The connection is that the representation of a polynomial by its value at an appropriate number of points is a modular representation. Consider

$$p(x)/(x - x_i) = q_i(x) + c_i/(x - x_i).$$

Then
$$p(x) = (x - x_i)q_i(x) + c_i,$$

so
$$p(x_i) = p(x) \bmod (x - x_i) = c_i.$$

Evaluation at many points thus corresponds to obtaining a modular representation, and exact interpolation corresponds to "recovering" the polynomial from its modular representation. The well-known fact that there exists a unique polynomial of degree $\leq n$ (i.e., unique mod $\prod_{i=0}^{n} (x - x_i)$), which fits the points $\langle x_0, y_0 \rangle, \ldots, \langle x_n, y_n \rangle$, is an immediate consequence of a generalized statement of the Chinese remainder theorem. In an abstract Euclidean domain setting (e.g., the ring of integers, the ring of polynomials), interpolation and Chinese remainder are the same process, although computationally there may be "tricks" applicable to one domain but not to the other. See Lipson (1971) for an excellent discussion of this correspondence. We will comment on some of the computational differences in later sections.

In the development of the FFT (Section 4.2) and for the general evaluation and interpolation problems (Section 4.5), we will use a well-known strategy, which might be called "divide and conquer," or "binary recursion." Informally, the strategy proceeds in the following manner: To solve a problem of "size" n (e.g., evaluating an $n - 1$ degree polynomial at n points):

(1) Split the problem into two subproblems of size $\sim (n/2)$.
(2) Recursively solve each subproblem.
(3) "Merge together" the solutions of the subproblems to form the solution of the given problem.

Probably the best known application of the strategy is Merge sort. If OVERHEAD (n) reflects the cost of steps 1 and 3, then the complexity of the procedure is described by the recurrence $T(n) = 2T(n/2) +$ OVERHEAD (n) If OVERHEAD (n) = $0(n \log^k n)$, then $T(n) = 0(n \log^{k+1} n)$ for any integer $k \geq 0$.

In Section 4.4, we will show that division (in either the polynomial or integer setting) has the same asymptotic complexity as multiplication. A strategy, which we might call "extrapolative recursion," yields the desired complexity analysis. Informally, to solve a problem of size n (or to precision n):

(1) Recursively solve a subproblem of size $n/2$ (solve to precision $n/2$).
(2) Extrapolate to the full solution (to full n precision).

If OVERHEAD (n) reflects the cost of step 2, the complexity is described by the recurrence of $T(n) = T(n/2) +$ OVERHEAD (n). For OVERHEAD (n) = $0(n \log^k n)$, $T(n)$ will also be $0(n \log^k n)$. The Ford Johnson (1959) recursive sort follows this strategy. More to our immediate concern, we suggest that Newton iteration be viewed as an iterative unwinding of extrapolative recursion. The quadratic convergence of Newton iteration informally tells us that if x is an n precision estimate for a root x_0 of $f(x)$, then $\phi(x) = x - f(x)/f'(x)$ will be (assuming convergence) a $2n$ precision estimate for x_0. In Section 4.4, Newton iteration will provide a basis for an asymptotically fast division algorithm.

4.2 The Discrete Fourier Transform

The discrete Fourier transform is to be viewed as a transformation from one *representation of a polynomial* to another representation.

The Discrete Fourier Transform

(a) Forward transform: {coefficient representation} → {point value representation}. That is, $\langle a_0, \ldots, a_n \rangle \to \langle\langle x_0, y_0 \rangle, \ldots, \langle x_s, y_s \rangle\rangle$, where $s \geq n + 1$.

(b) Inverse transform (interpolation): {point value representation} → {coefficient representation}.

We now investigate how to perform a *fast* forward and inverse transform. If we had an $O(n \log n)$ algorithm for these transforms, then we could do polynomial multiplication $(p(x) = \Sigma a_j x^j) \times (q(x) = \Sigma b_j x^i)$ in $F[\vec{a}, \vec{b}]$, given $F \cup \{\vec{a}, \vec{b}\}$, as follows:

(1) Evaluate $\left.\begin{array}{l} p(x)|_{x=x_i} \\ \\ q(x)|_{x=x_i} \end{array}\right\}$ $0 \leq i \leq 2n$
$\deg p = \deg q = n; \{x_i\} \subseteq F$

(2) Evaluate $w(x_i) = p(x_i) \times q(x_i)$, $0 \leq i \leq 2n$.

(3) Interpolate $\{\langle x_i, w(x_i)\rangle\}$ to get (the coefficients of) the unique polynomial of degree $\leq 2n$ which satisfies this set of points.

The costs are:

(1) $2\, O(2n \log 2n) = O(n \log n)$ scalar operations;
(2) $2n + 1$ nonscalar multiplications;
(3) $O(2n \log 2n) = O(n \log n)$ scalar operations;
   total operations = $O(n \log n)$; nonscalar multiplications = $O(n)$.

Classically, if we evaluate an nth degree polynomial at $n + 1$ points (say, by Horner's rule), we use $2n(n + 1) = O(n^2)$ operations. Similarly, classical interpolation (say, by Lagrange's or Newton's method) uses $O(n^2)$ operations. We will return to the general evaluation and interpolation problem in Section 4.5, but for present purposes it is important to note that *any set of evaluation points* is permissible (although there are many "modular problems" where certain primes or moduli may be "bad"). Can we choose the points of evaluation $\{x_i\}$ in such a way as to enable fast transformations? (Of course. We have already answered this question.)

DEFINITION 4.2.1

ω is a *primitive kth root of unity* if $\omega^k = 1$ and $\omega^j \neq 1, \forall j < k$. For the following development, assume we are working in the field of complex numbers. For example,

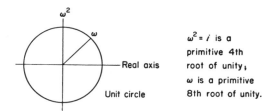

The following property is basic to the development of the FFT.

LEMMA 4.2.2

Let ω be a primitive $(n + 1)$st root of unity. Then

$$\sum_{0 \leq s \leq n} \omega^{s\alpha} = \begin{cases} n + 1 & \text{if } \alpha = \mod (n + 1) \\ 0 & \text{otherwise} \end{cases}$$

Proof. If $\alpha = 0 \mod (n + 1)$, then $\omega^{s\alpha} = 1$ by definition. Suppose $\alpha \neq 0 \mod (n + 1)$. Then

$$\sum_{0 \leq s \leq n} \omega^{s\alpha} = \sum_{s=0}^{n} z^s \Big|_{z = \omega^\alpha}$$

$$= \frac{z^{n+1} - 1}{z - 1} \Big|_{z = \omega^\alpha} \qquad \text{(This holds in any field or integral domain, since } z = \omega^\alpha \neq 1\text{)}$$

$$= 0 \qquad \text{(since } z^{n+1} = \omega^{\alpha(n+1)} = 1\text{)}$$

□

And now the FFT.

THEOREM 4.2.3. (See Cooley, Lewis, and Welch [1967] for a history of the fast transform)

The choice of points $x_i = \omega^i$, $0 \leq i \leq n$, where ω is a primitive $(n + 1)$st root of unity, allows us to compute the forward and in-

The Discrete Fourier Transform

verse transform of an arbitrary nth degree polynomial in $0(n \log n)$ operations.

Proof. We want to show why the forward transform and its inverse are fast, using this set of points.

(a) Evaluation (forward) transform:

$$p(x) = \sum_{i=0}^{n} a_i x^i \bigg|_{x=x_j}$$

Assume $n + 1 = 2^r$. (This does not affect the asymptotic complexity, but it is bothersome in certain applications.)

$$p(x) = a_0 + a_1 x + a_2 x^2 + \cdots a_n x^n$$

$$= (a_0 + a_2 x^2 + \cdots a_{n-1} x^{n-1}) + (a_1 x + a_3 x^3 + \cdots a_n x^n)$$

$$= (a_0 + a_2 x^2 + \cdots a_{n-1} x^{n-1}) + x(a_1 + a_3 x^2 + a_n x^{n-1}).$$

Substitute: $y = x^2$

$$= (a_0 + a_2 y + \cdots a_{n-1} y^{(n-1)/2})$$

$$+ x(a_1 + a_3 y + \cdots a_n y^{(n-1)/2})$$

$$= p_1(y) + x p_2(y),$$

where $\deg p_1 = \deg p_2 = (n-1)/2 = 2^{r-1} - 1$. Evaluate at $x = x_0, \ldots, x_n$; i.e., evaluate p_1 and p_2 at $y = y_0, \ldots, y_n$, where $y_i = x_i^2$. For example, $n = 7$:

ω is an 8th root of unity $\quad x_0 = 1$
$\qquad\qquad\qquad\qquad\qquad x_1 = \omega$
$\qquad\qquad\qquad\qquad\qquad x_i = \omega^i$

$y_0 = 1$
$y_1 = \omega^2$
$y_2 = \omega^4$
$y_3 = \omega^6$

$y_4 = \omega^8 = 1$
$y_5 = \omega^{10} = \omega^2$
$y_6 = \omega^{12} = \omega^4$
$y_7 = \omega^{14} = \omega^6$

To evaluate at $y = y_i$
$0 \leq i \leq 7$
is to evaluate at $y = y_i$
$0 \leq i \leq 3$

This process will continue to work (recursively) because we started with ω as a kth root of unity $k = n + 1 = 2^r$, and hence ω^2 is a $(k/2)$th root of unity. Let's look at the cost $T(2^r)$ = number of arithmetics for evaluating $\sum_{i=0}^{2^r-1} a_i x^i$ at 2^r selected points $\{\omega^j \mid 0 \leq j \leq 2^r - 1\}$,

where ω is a (2^r)th root of unity. $T(2^r) = 2T(2^{r-1}) + 2 \cdot 2^r$, where the last term represents one + and one × for each x_i. We can cut the number of multiplications in half by recognizing that $x_{i+2^{r-1}} = -x_i$ and then we have the recurrence

$$T(2^r) = 2T(2^{r-1}) + (3/2) \cdot 2^r$$

$$T(1) = 0$$

∴ $T(2^r) = (3/2) \cdot 2^r \cdot r$ and $T(n) = 0(n \log n)$ when $r = \lceil \log n \rceil$.

(b) Inverse transformation (interpolation)

$$\{\langle x_i, y_i \rangle\} \to \{a_i\}$$

where $x_i = \omega^i$, ω an $(n+1)$st root of unity.
View evaluation as a vector-matrix multiplication.

$$(a_0, \ldots, a_n) \begin{pmatrix} 1 & 1 & 1 & \cdots & 1 \\ 1 & \omega & \omega^2 & \cdots & \omega^n \\ 1 & \omega^2 & \omega^4 & \cdots & \cdots \\ \vdots & & & & \\ 1 & \omega^n & \omega^{2n} & & \omega^{n^2} \end{pmatrix} = (y_0, \ldots, y_n)$$

If V is the above Vandermonde matrix, then $(a_0, \ldots, a_n) = (y_0, \ldots, y_n) V^{-1}$.

The Discrete Fourier Transform

Define $\tilde{V} = (\tilde{v}_{i,j})$ by $\tilde{v}_{i,j} = \omega^{-ij}/(n+1)$. We will show $V^{-1} = \tilde{V}$.

$$(V\tilde{V})_{i,j} = \frac{\sum_{0 \leq s \leq n} \omega^{is} \omega^{-js}}{(n+1)}$$

$$= \frac{\sum_{0 \leq s \leq n} \omega^{(i-j)s}}{(n+1)}$$

Now use Lemma 4.2.2 to verify that

$$(V\tilde{V})_{i,j} = \begin{cases} 1 & i=j \\ 0 & i \neq j \end{cases}$$

and hence $\tilde{V}$ is indeed the inverse of V. Then

$$(a_0, \ldots, a_n) = (y_0, \ldots, y_n) V^{-1}$$

$$= (y_0, \ldots, y_n) \begin{pmatrix} 1 & 1 & \cdots & 1 \\ 1 & \omega^{-1} & & \omega^{-n} \\ \vdots & & & \\ 1 & \omega^{-n} & & \omega^{-n^2} \end{pmatrix} \Big/ (n+1)$$

$$= \sum_{i=0}^{n} y_i z^i \Big|_{\substack{z=\omega^{-i} \\ 0 \leq i \leq n}} \Big/ (n+1)$$

But if ω is an $(n+1)$st root of unity, then so is ω^{-1}. So, we have shown that interpolation with respect to the "Fourier points" is equivalent to doing a forward transformation and hence needs only $O(n \log n)$ arithmetics. □

Note that only (complex) scalar multiplications and ± operations were used for the transforms.

COROLLARY 4.2.4 (polynomial multiplication)

Let

$$u(x) = \sum_{i=0}^{n} u_i x^i, \quad v(x) = \sum_{i=0}^{n} v_i x^i$$

and

$$w(x) = u(x) \times v(x) \stackrel{\circ}{=} \sum_{i=0}^{2n} w_i x^i.$$

We can compute $w_0, \ldots, w_{2n}$ in $C[u_0, \ldots, u_n, v_0, \ldots, v_n]$ given $C \cup \{u_i\} \cup \{v_i\}$ by a program that uses only $2n + 1$ nonscalar operations and $O(n \log n)$ total operations.

How practical is this method? While relatively few arithmetics are needed ($\sim 9n \log n$), the operations are in the complex field and a recursive implementation, while conceptually neat, might execute very slowly (since the stacking mechanism in recursive procedures is usually software implemented rather than hardware). J. Lipson (1974), G. Sande, and others have implemented fast transforms. The crossover points (as compared to classical evaluation and interpolation) for a recursive implementation seems to be around degree 100 (Lipson). Sande's implementation, which is not based on the indicated recursion and which is more "hand tailored" for different degrees n, may be advantageous for degrees as small as 25.

What about exact computation for integer polynomials? It is possible in some applications to make the complex arithmetic sufficiently accurate so that exact integer values can be recovered. A "neater" approach, however, is to perform the transform over a finite field (rather than the field of complex numbers). Again, let us assume that $n + 1 = 2^r$, where n is the degree of the polynomial being transformed.

FACT 4.2.5

If $M = c \cdot 2^r + 1$ is a prime, then the field Z_M has a primitive (2^r)th root of unity.

The proof follows from the basic result (in algebra or number theory) that for M prime, Z_M has a primitive kth root of unity if and only if k divides $M - 1$ (see Hardy and Wright [1960].) For example, let $M = 5 \cdot 2^3 + 1 = 41$. Then $\omega = 3$ is a primitive ($2^3 = 8$)th root of unity in Z_{41}. Hence, we can carry out the arithmetic (for the transform) mod M, at a classical cost of $0(\log M)^2$ bit operations (single precision operations) per arithmetic operation mod M. Thus, if M is too large, we might defeat any savings gained by a fast transform.

Depending on the application, there will be a lower bound on the size of M. For example, in polynomial multiplication, M must be $\geq$ the largest coefficient of the product polynomial (which can be bounded by $(n+1)b^2$, where b is the largest coefficient in the given polynomials). From a practical point of view, Lipson (1974) has shown that a "good" choice of M is always obtainable. Specifically, he has shown that word-size primes M of the form $c2^r + 1$ are quite dense. If the target (i.e., product) coefficients are larger than word size, then modular techniques can be used with the assurance that there will be a sufficient number of primes for any practical application. (There are about 50 primes $M < 2^{31}$ of the form $c2^r + 1$ where $r \geq 20$; thus, we can expect to easily handle polynomials of degree $n \leq 2^{20} \approx 10^6$.)

For many (most?) applications, the cost of the transform does not dominate the complexity. Consider the solution of a set of linear equations having symbolic coefficients. If the matrix is n by n and the symbolic coefficients are degree-d univariate polynomials, then the determinant will have degree $\leq nd$. We are assuming that the solution has polynomial rather than rational function entries or that we are using a fraction-free variant of Gaussian elimination (see Lipson [1968]). Thus, straightforward Gaussian elimination will result in $0(n^3)$ "steps," where each step requires (classically) $\leq 0(nd)^2$ arithmetic operations; the total number of arithmetic operations then is $0(n^5 d^2)$. Even if the "steps" are performed with a fast transform, the total cost would be $0(n^4 d(\log nd))$.

Alternatively, we can transform all polynomials immediately (assuming them to be of degree nd), carry out Gaussian elimination on each of the $nd + 1$ transform values, and then interpolate back. The total cost is the cost of transform (for degree nd) plus $0(nd)n^3$, i.e., $0(n^4 d)$, even if we use classical evaluation and interpolation for the transforms.

Finally, we should note that modular techniques may be quite inefficient when the input polynomials are "sparse." In practice, do we "usually" encounter only sparse polynomials or polynomials of small degree? It is this point of view that distinguishes most algebraic and symbolic manipulation systems (e.g., see Hall [1971] for a discussion of the ALTRAN system) from Collin's (1971) SAC system.

4.3 Fast Number Multiplication

For a number of years several people had sensed the close relationship between fast n-degree polynomial multiplication and n-bit number multiplication. But it was not until recently that Schönhage and Strassen (1971) demonstrated a way by which the analogy could be exploited, yielding an $O(n \log n \log \log n)$ method. The Schönhage-Strassen method is presently the fastest algorithm in the asymptotic sense, and the fact that a bound near $O(n \log n)$ can be obtained is remarkable. (However, it is not at all clear that the method could ever be practically implemented.) While the basic ideas are all comprehensible, the details of the algorithm are sufficiently formidable that we choose instead to present a simplified $O(n \log^2 n)$ version given by Karp (1971) in his lecture notes. For motivation, we will first present a nonasymptotic algorithm suggested by Pollard [1971] and analyzed by Lipson [1974].

The Pollard-Lipson algorithm will work only for $n \leq N$, where N is large enough to handle most "foreseeable applications." For its domain, the algorithm has complexity (number of bit operations or number of single precision operations) $\leq cn \log n$ (for a "reasonable" constant c).

FACT 4.3.1

For the fast transform, with respect to exact multiplication of integer polynomials, we can work in a finite *ring* Z_M if it has the appropriate properties ($r = 2^s$ = degree required by the application):

(1) There is an rth root of unity ω in Z_M.
(2) r has an inverse in Z_M.
(3) $\displaystyle\sum_{0 \leq s \leq r-1} \omega^{s\alpha} = \begin{cases} r & \text{if } \alpha = 0 \bmod r \\ 0 & \text{if } \alpha \neq 0 \bmod r \end{cases}$

The fact follows from the constructions in the fast transform. We are now ready to describe a "practically asymptotic" fast number multiplication algorithm.

Fast Number Multiplication

THEOREM 4.3.2

Let u and v be K precision numbers. (There will be a bound on how large K can be.) Then $u \times v$ can be computed in $O(K \log K)$ single precision operations. (Recall, in contrast, that the "schoolboy" method is $O(K^2)$.)

Proof. The main idea is to view both numbers, u and v, as polynomials, $u(x)$ and $v(x)$, evaluated at a specific point $x = 2^L$. So, $u \times v$ is the product polynomial evaluated at the same point. Let u and v be $N = KL$ bit numbers broken up into K pieces, L bits in each piece. For example,

$$\begin{array}{ccccc|ccccc} a_4 & a_3 & a_2 & a_1 & a_0 & b_4 & b_3 & b_2 & b_1 & b_0 \\ u = \ 101 & 001 & 010 & 000 & 100 & v = \ 000 & 111 & 011 & 111 & 000 \end{array}$$

Then

$$L = 3 = \text{number of bits in each piece};$$
$$K = 5 = \text{number of pieces};$$

$$u = \sum_{i=0}^{K-1} a_i x^i \bigg|_{x=2^L} = u(x) \bigg|_{x=2^L} ;$$

$$v = \sum_{i=0}^{K-1} b_i x^i \bigg|_{x=2^L} = v(x) \bigg|_{x=2^L}$$

Let $w(x) = u(x) \times v(x)$. Then

$$w = u \times v = u(x) \times v(x) \bigg|_{x=2^L} = w(x) \bigg|_{x=2^L}$$

$$= \sum_{i=0}^{2K-2} w_i x^i \bigg|_{x=2^L}$$

In the Schönhage-Strassen method, $K \approx L \approx \sqrt{N}$ and we work in a finite ring satisfying Fact 4.3.1. But for now choose L to be word size (or maybe half-word size). Note that this choice of L lends itself to practical considerations of I/O presentation as well as the full utilization of single precision arithmetic.

The big difference between polynomial multiplication and number multiplication is that there is no carry in the former case. That is, when we form the coefficients of $w(x)$, we do not have appropriate sections of the number w because the carry propagation has been inhibited. We can "recover" w from $w(x)$ by "releasing the carry," which is equivalent to doing the evaluation $w(x)\big|_{x=2^L}$.

Now it only remains to analyze the cost of the polynomial multiplication and the subsequent evaluation or carry release. The maximum value of the coefficients of $w(x)$ is $\leq Kb^2$ ($b = 2^L$). Assume (and here is where we rule out an asymptotic method) that $2K = 2^r \leq b$, and hence each coefficient w_j is less than $b^3/2$. Choose three word-size primes:

$$p_i: \prod_{j=1}^{3} p_i \geq b^3/2, \quad p_i = c_i 2^r + 1.$$

(1) Compute the coefficients $w_j^i = w_j \bmod p_i$ of $w(x)$ by the FFT. Since the transform is performed at $2K$ points, the number of arithmetics required is

$$3 \cdot \begin{cases} 3/2 \ (2K \log K) & \text{(forward transform)} \\ + 2K & \text{(multiplication)} \\ + 3/2 \ (2K \log 2K) & \text{(inverse transform)} \end{cases}$$

or about $18K \log K$ arithmetics *modulo one of the p_i*. In practice we may need an equal number of "remainder operations," yielding a total of $36K \log K$ single precision arithmetics.

(2) Reconstruct the w_j from the w_j^i at a cost of $O(K)$.

(3) Release the carry. (Note that we have $2K - 1$ coefficients and the required number has $\leq 2K$ single precision pieces; hence the carry cannot propagate too far.

$$\tilde{w}_0 = \text{remainder } [w_0/b]$$

$$\tilde{w}_1 = \text{remainder } [(w_1 + \lfloor w_0/b \rfloor)b]$$

$$\vdots \qquad \vdots$$

$$\tilde{w}_K = \text{carry from last division by } b$$

It is not too hard to show that the numbers being divided do not get too big.

Fast Number Multiplication

$$w_0 \leq b^3$$

$$w_1 + \lfloor w_0/b \rfloor \leq b^3 + b^2$$

$$w_2 + \lfloor (b^3 + b^2)/b \rfloor \leq b^3 + b^2 + b^1$$

$$\vdots$$

$$w_{K-1} + \lfloor \quad \rfloor \leq 2b^3$$

Hence, the cost of releasing the carry $\sim 3K$. The total cost, then, to multiply two $N = KL$ bit numbers ($K \leq 2^L$) is $O(K \log K) \approx 36K \log K$ single precision operations.

On the IBM 370, $N \leq 2^{31} \cdot 31$ bits ≈ 40 billion decimal digits. It is in this sense that it is appropriate to call the Pollard-Lipson method "practically asymptotic." With a good implementation this method becomes superior to the classical method for numbers having around 500 decimal digits. By using one prime and further restricting the input size, we may lower the "crossover point" to numbers having 100 decimal digits.

We are now ready to present Karp's simplified version of the Schönhage-Strassen algorithm. We again view u and v as $N = K \cdot L$ bit numbers, but now $K \approx L$. We can assume $K = 2^k$, $L = 2^l$, $N = 2^n$ with $n = k + l$. The product is $w = \sum_{i=0}^{2K-2} w_i x^i \big|_{x=2^L}$ and we know that $0 \leq w_i \leq K \cdot 2^{2L}$. Since $2^{2L} + 1$ and $2^k = K$ are relatively prime, it follows (by the Chinese remainder theorem) that every w_i is uniquely determined by $w_i \bmod K$ and $w_i \bmod (2^{2L} + 1)$. So here is the method:

(1) Compute $\{w_i \bmod K \mid 0 \leq i \leq 2K - 2\}$.
(2) Compute $\{w_i \bmod 2^{2L} + 1 \mid 0 \leq i \leq 2K - 2\}$.
(3) Compute $\{w_i \mid 0 \leq i \leq 2K - 2\}$.
(4) Release the carry to compute $w = u \times v$.

The limiting computation turns out to be step 2. It can be shown (see problems 4.3 and 4.4) that the cost $T_1(N)$, in bit operations, for steps 1, 3, and 4 is $O(N)$.

We want to compute $\{w_i \bmod 2^{2L} + 1 \mid 0 \leq i \leq 2K - 2\}$, using a fast transform relative to the finite ring $Z_{2^{2L}+1}$. According to Fact 4.3.1, we need to find an $r = 2^s \leq 2K - 1$ and an ω satisfying:

(1) $\omega^r = 1 \bmod (2^{2L} + 1)$ and $\omega^t \neq 1$ for $t < r$. It suffices to have $\omega^{r/2} = -1 \pmod{2^{2L} + 1}$; i.e., we can choose ω and r so that $\omega^{r/2} = 2^{2L}$ and therefore (assuming r divides $4L$) that $\omega = 2^{4L/r}$. If we set $r = 2K = 2^{k+1}$, then $\omega = 2^{2L/K}$, assuming K divides $2L$; i.e., assuming $k \leq l + 1$. We claim

(2) $\displaystyle\sum_{0 \leq s < r} \omega^{s\alpha} = \begin{cases} r & \text{if } \alpha = 0 \bmod r \\ 0 & \text{otherwise} \end{cases}$

(3) $r^{-1} \bmod (2^{2L} + 1)$ exists.

So, by Fact 4.3.1 and the analysis of Theorem 4.2.3, we can compute $\{w_i \bmod 2^{2L} + 1\}$ in $T_2(N)$ bit operations, where $T_2(N)$ is the sum of the following costs:

(a) Forward transform: $O(K \log K)$ operations in $Z_{2^{2L}+1}$, each of which is either an addition or a multiplication with one operand being $\omega^{ij} = 2^t$ for some t.
(b) Multiplication of transformed polynomials: $2K$ general multiplications in $Z_{2^{2L}+1}$
(c) Inverse transform: $O(K \log K)$ operations as in (a).

To further analyze these costs, look at arithmetic in a ring Z_{2^p+1}. We can express the numbers in Z_{2^p+1} using p bits if we do not express $-1 \equiv 2^p$. Addition and multiplication involving -1 is a special but easy case. Indeed, multiplication by any 2^t in Z_{2^p+1} can be done in $O(p)$ bit operations by shifting t places and then subtracting the carry (= the high-order bits). For example, in $Z_{2^3+1} = Z_9$ we have

Fast Number Multiplication

$$\begin{array}{r} 101 \\ 100 \\ \underline{10} \quad 100 \\ \hookrightarrow \underline{10} \\ 010 \end{array} \qquad \begin{array}{r} 5 \\ \underline{4} \\ 20 = 2 \bmod 9 \end{array}$$

A general multiplication in Z_{2^p+1} of numbers $u_1, v_1 < 2^p$ is simply a multiplication of two p bit numbers with the appropriate carry applied to the low-order p bits of interest. We apply the complete algorithm recursively to compute this multiplication.

We can now express the number of bit operations $T(N)$ for the complete algorithm as $T(N) = T_1(N) + T_2(N)$, where $T_1(N) = 0\,(N)$ and $T_2(N)$ is the sum of the following costs:

(a) $O(K \log K) \cdot O(L) = (KL \log K)$;
(b) $2K \cdot T(2L)$;
(c) $O(KL \log K)$.

Note that $T_1(N)$ is absorbed in the cost for $T_2(N)$. We want to optimize $K = 2^k$ and $L = 2^l$ subject to $KL = N = 2^n$ and $k \leq l + 1$.

$$T(N) = O(KL \log K) + 2K \cdot T(2L);$$

$$\frac{T(N)}{N} = \frac{O(KL \log K)}{KL} + \frac{2K \cdot T(2L)}{KL}$$

$$= O(\log K) + 2^2 \cdot \frac{T(2L)}{2L}.$$

Let

$$F(n) = \frac{T(2^n)}{2^n} = \frac{T(N)}{N}.$$

Then

$$F(n) = O(k) + 2^2 F(l+1).$$

We would like to make l as small as possible while satisfying $k \leq l + 1$. Therefore, choose $l = \lfloor n/2 \rfloor$, $k = \lceil n/2 \rceil$. It follows that

$$F(n) = O(n^2),$$

and so $\quad T(N) = N \cdot F(n) = N(\log N)^2.$

□

We can summarize these ideas in the surprising result:

THEOREM 4.3.3 (Schönhage and Strassen [1971].)

Two N precision numbers (N bit numbers) can be multiplied in $M(N) = O(N \log N \log \log N)$ single precision operations (bit operations or steps on a multitape Turing machine).

In order to get the Schönhage-Strassen bound of $N \log N \log \log N$, we would like the recurrence in the preceding analysis to be $F(n) = O(k) + 2F(l + 1)$. The idea is to develop an algorithm to compute $u \cdot v \bmod (2^N + 1)$ which will use only K recursive calls of multiplications $\bmod (2^{2L} + 1)$. The substantial details can also be found in Knuth (1973b).

4.4 Fast Integer and Polynomial Division

In any Euclidean domain E, the concept of division is meaningful, for by definition (see van der Waerden [1964]), there is a degree function $d : E - \{0\} \to \mathbf{R}$ satisfying the property

$$\forall\, u, v \neq 0, \quad \exists\, q, r : u = qv + r \quad \text{and} \quad d(r) < d(v)$$

For $E = F[x]$ and $d(u) = \deg(u)$ we know that q and r are uniquely defined. For $E = \mathbf{Z}$ and $d(u) = \log_b |u|$ for any base b, or for the more familiar $d(u) = |u|$, we need the added condition $r \geq 0$ for uniqueness. Let

$$\text{Precision } u \triangleq \begin{cases} \deg u + 1 & \text{for } u \in F[x] \\ \lfloor \log_b |u| \rfloor + 1 & \text{for } u \in I \end{cases}$$

(A more general discussion of precision and step in the context of Euclidean domains can be found in Moenck [1973] and Moenck and Allen [1975].) Our goal is to show that the cost $D(n)$ of computing

Fast Integer and Polynomial Division

q and r (from v and n precision u) is proportional to $M(n)$, the cost of multiplying two n precision elements. It suffices to show how to compute q, since r can then be obtained in one additional multiplication and subtraction.

It is interesting that integer division was reduced to multiplication before the same reduction became known for the polynomial setting. That is somewhat understandable, given that interest in software arithmetic precedes symbolic polynomial computation, and also that the role of the reciprocal $1/v$ may be more apparent in the integer case. But with hindsight, we will see that the polynomial setting constitutes a simpler computational problem. Moenck and Borodin (1972) first gave an $O(M(n) \log n) = O(n \log^2 n)$ algorithm; Strassen (1972) soon improved this to $O(M(n))$ by showing how to use Sieveking's (1972) power series reciprocal. We follow Kung's (1973c) exposition for the reciprocal computation.

THEOREM 4.4.1

Let $u = \sum_{i=0}^{n} u_i x^i$, $v = \sum_{i=0}^{m} v_i x^i$ be in $F[x]$. We can compute (the coefficients of) the unique $q(x)$ and $r(x)$ satisfying $u = qv + r$, $\deg r < \deg v$ in $O(M(n)) = O(n \log n)$ arithmetic steps (assuming F supports a fast transform) and $O(n)$ nonscalar operations. The computation takes place in $F(u_0, \ldots, u_n, v_0, \ldots, v_m)$, given $F \cup \{u_i\} \cup \{v_j\}$.

Proof. A few observations will reduce the problem to that of computing a power series reciprocal. We assume $m \leq n$.

(a) It suffices to compute q, since r can then be obtained in one additional polynomial multiplication and subtraction (i.e., cost $O(M(n))$.
(b) We can assume $v_m = 1$ (i.e., can divide u and v by v_m at cost of $O(m + n)$). With $v_m = 1$, the computation is in $F[\vec{u}, \vec{v}]$; i.e., division is not needed.
(c) $u(x) = q(x)v(x) + r(x)$. Let $z = 1/x$, then

$$\sum_{i=0}^{n} u_i z^{-i} = \sum_{i=0}^{n-m} q_i z^{-i} \cdot \sum_{i=0}^{m} v_i z^{-i} + \sum_{i=0}^{m-1} r_i z^{-i}$$

multiplying the equation by z^n, we get

$$\sum_{i=0}^{n} u_{n-i} z^i = \sum_{i=0}^{n-m} q_{n-m-i} z^i \cdot \sum_{i=0}^{m} v_{m-i} z^i$$

$$+ \sum_{i=0}^{m-1} r_{m-1-i} z^{n-m+1+i}$$

and therefore (expressing the equality mod z^{n-m+1}),

$$\sum_{i=0}^{n-m} u_{n-i} z^i = \sum_{i=0}^{n-m} q_{n-m-i} z^i \cdot \sum_{i=0}^{n-m} v_{m-i} z^i \quad (\bmod\ z^{n-m+1})$$

It is tempting to complete the proof of the theorem now, for we can apply the transform or modular concept here, evaluating

$$\sum_{i=0}^{n-m} q_{n-m-i} z^i = \frac{\sum u_{n-i} z^i}{\sum v_{m-i} z^i}$$

at $n - m + 1$ points. Computationally, we must avoid "bad" moduli—those evaluation points that are zeros of the denominator. Moreover, we have not reduced polynomial division to polynomial multiplication as indicated in the theorem.

(d) Since $v_m = 1$, $\sum_{i=0}^{n-m} v_{m-i} z^i$ has an inverse g in the power series ring $F[[z]]$. That is, $(\sum_{i=0}^{n-m} v_{m-i} z^i) \cdot g = 1$ and therefore

$$(\sum_{i=0}^{n-m} u_{n-i} z^i) g = \sum_{i=0}^{n-m} q_{n-m-i} z^i \quad (\bmod\ z^{n-m+1})$$

This implies

$$(\sum_{i=0}^{n-m} u_{n-i} z^i) \cdot (g\ \bmod\ z^{n-m+1}) = \sum_{i=0}^{n-m} q_{n-m-i} z^i \quad (\bmod\ z^{n-m+1})$$

Thus, it suffices to be able to compute the polynomial $g\ \bmod\ z^{n-m+1}$, which we can view as an approximation to the power series g, the reciprocal of $\sum v_{m-i} z^i$.

Our goal, then, is to be able to compute better and better (= higher and higher degree) approximations to a power series reciprocal.

Fast Integer and Polynomial Division

Let $\alpha = \sum_{i=0}^{\infty} \alpha_i z^i \in F[[z]]$ be an arbitrary power series with $\alpha_0 \neq 0$ (in particular, α could be a polynomial), and let $\beta = \sum_{i=0}^{\infty} \beta_i z^i$ satisfy $\alpha\beta = 1$. We need to know only α mod z^N to compute β mod $z^N = \sum_{i=0}^{N-1} \beta_i z^i$. The extrapolative recursion strategy suggests that we want to compute the N precision β mod z^N and then extrapolate to the $2N$ precision β mod z^{2N}. We use the words "precision" and "approximation" to suggest the use of Newton iteration, which may be thought of as an iterative realization of an extrapolative recursion.

In the present situation we want to compute β, the root of $f(y) = (1/y) - \alpha$. The Newton iteration is $\phi(y) = y - f(y)/f'(y) = 2y - \alpha y^2$. We want to show that this situation is "converging" properly.

$$\phi(y) - \beta = 2y - \alpha y^2 - \beta$$
$$= y - \beta + y - \alpha y^2$$
$$= y - \beta - ((y/\beta)(y - \beta))$$
$$= (1 - (y/\beta))(y - \beta)$$
$$= -\alpha(y - \beta)^2$$

Thus, if $y - \beta = 0$ mod z^N, then $\phi(y) - \beta = 0$ mod z^{2N}. Finally, the (extrapolation) computation $\phi(z) = 2z - \alpha z^2 = z(2 - \alpha z)$ entails two $2N$ precision multiplications, and therefore $T(2N) = T(N) + 0(M(N)) = 0(M(N))$. □

We now return to the integer setting, where for $v \in \mathbf{N}$ it is apparent what $1/v$ means; it is a fraction, which we can represent to the accuracy desired as $(.w_1 \cdots w_r)_b$. As in Theorem 4.4.1, the computation of the reciprocal $1/v$ will be accomplished using Newton iteration, but now more care is needed to insure that we converge properly.

THEOREM 4.4.2 (Cook; see Knuth [1969], p. 275].)

Given n precision (bit) u and m precision (bit) v in $\mathbf{N}$, we can compute the unique q and r such that $u = qv + r$, $0 \leqslant r < v$, using $0(M(n)) = 0(n \log n)$ single precision steps (multitape Turing machine steps).

Proof. Let $u = (u_1 \cdots u_n)_2$, $v = (v_1 \cdots v_n)_2$. As in Theorem 4.4.1, a few observations reduce the problem to that of computing the reciprocal.

(a) It suffices to compute q.

(b) Multiplying the equation $u = qv + r$ by 2^{-m}, we obtain $\tilde{u} = q\tilde{v} + \tilde{r}$ with $\tilde{u} = u_1 \cdots u_{n-m} \cdot u_{n-m+1} \cdots u_n$, $\tilde{v} = .v_1 \cdots v_m$ and $0 \leq \tilde{r} < \tilde{v}$.

(c) Suppose we can compute z satisfying $|(1/\tilde{v}) - z| \leq 2^{-(n-m)}$. Then $\hat{q} \stackrel{\circ}{=} \tilde{u}z$ satisfies $|\hat{q} - q| \leq 2$. We can correct $\hat{q}$ to find q so that $0 \leq u - qv < v$.

The reciprocal approximation (i.e., z) is computed essentially by the Newton iteration $y_{i+1} = \phi(y_i) = 2y_i - \tilde{v}y_i^2$. We want to perform this iterative computation with increasing precision so that if y_i satisfies $0 \leq (1/\tilde{v}) - y_i \leq 2^{(-2^{i-1}+1)}$, then y_{i+1} satisfies $0 \leq (1/\tilde{v}) - y_{i+1} \leq 2^i + 1$).

Then $T(2^i)$, the cost to obtain an approximation z with $|(1/\tilde{v}) - z| \leq 2^{-2^i}$, satisfies the recurrence $T(2^{i+1}) = T(2^i) + O(M(2^i))$, and hence, $T(2^i) = O(M(2^i))$. It can be shown (see Knuth [1969], p. 275) that the desired convergence can be achieved by computing $\hat{y}_{i+1} \leftarrow 2y_i - (.v_1 \cdots v_{2^{i+1}+3})y_i^2$ and then rounding up $\hat{y}_{i+1}$ to y_{i+1} so that y_{i+1} is a multiple of $2^{(-2^{i+1}-1)}$. Note that the complexity of this step is dominated by the two multiplications, $y_i \times y_i$ and $(.v_1 \cdots v_{2^{i+1}+3})y_i^2$.

□

4.5 Fast Evaluation and Interpolation
(Fast Modular Representation and Chinese Remainder Algorithms)

The previous results suggest a number of questions. We have already seen in Section 2.6 that many-point polynomial evaluation (i.e., evaluating a degree $n - 1$ polynomial at n points) can be reduced from $O(n^2)$ by fast matrix multiplication, but only to $O(n^{1.91})$. Can this be improved to something near $O(n \log n)$, like the evaluation at the special Fourier points? What is the asymptotic difficulty of the inverse problem, general interpolation? And in the dual integer setting, how fast can we obtain the (single precision) modular

Fast Evaluation and Interpolation

representation of an n precision number and how fast can we perform a Chinese remainder process?

We want to exploit the divide and conquer strategy (see Section 4.1) to obtain fast evaluation and interpolation algorithms. To lay bare the algebraic nature of the results, we will follow the development in Borodin and Moenck (1974). The results are based on the earlier works of Fiduccia (1972 b), Horowitz (1972), and Moenck and Borodin (1972). We are interested in modular forms (evaluation) and Chinese remainder (interpolation) algorithms for a Euclidean domain E satisfying the additional assumptions that

(a) $M(n) \triangleq$ "cost of multiplying two n precision numbers u, v of E" is $O(n \log n)$; and
(b) $D(n) \triangleq$ "cost of dividing u of precision n by v of precision $m \leqslant n$" is $O(M(n))$.

We are, admittedly, motivated by (and perhaps only interested in) the integer domain $\mathbf{Z}$ and the polynomial domain $F[x]$, F a field. For $u \neq 0$ in $\mathbf{Z}$, precision $(u) = \lfloor \log_b |u| \rfloor + 1$ relative to any base b; for $u \in F[x]$, precision u is $\deg u + 1$. Previous sections establish assumptions (a) and (b) for these domains. (To unify the discussion we are assuming $M(n)$ in $\mathbf{Z}$ is $O(n \log n)$ rather than $O(n \log n \log \log n)$.) For those who are justifiably skeptical about the claimed generality, we note that the analysis also applies to the Gaussian integers.

For definiteness and to further motivate the results in Chapter 5, we will state explicit corollaries relative to the polynomial setting. The bounds on total arithmetic operations in these corollaries assume F supports a fast transform. The nonscalar operation bounds are not dependent on F, since any set of integers could be used for the transform, simulating multiplication with repeated additions (assuming char $F = 0$). Without loss of generality, the discussion is simplified by assuming $n = 2^r$.

FACT 4.5.1

Let $m_1, \ldots, m_n$ be single precision. We can compute the "super-moduli": $m_1 m_2, \ldots, m_{n-1} m_n, \prod_{i=1}^{4} m_i, \ldots, \prod_{i=n-3}^{n} m_i, \ldots, \prod_{i=1}^{n} m_i$, using only $O(M(n) \log n) = O(n \log^2 n)$ steps.

Proof. The computation tree shown below is just an unwinding of the binary recursion based on the equality $\prod_{i=1}^{n} m_i = \prod_{i=1}^{n/2} m_i \cdot \prod_{n/2+1}^{n} m_i$.

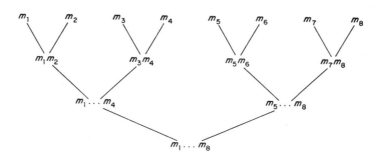

In terms of the polynomial setting we have the following corollary.

COROLLARY 4.5.2

In $O(M(n) \log n)$ we can compute (the coefficients of) the polynomials $(x - x_1)(x - x_2), \ldots, (x - x_{n-1})(x - x_n), \ldots, \prod_{i=1}^{n} (x - x_i)$. In particular, we can compute the elementary symmetric functions $\{p_1(x_1, \ldots, x_n), \ldots, p_n(x_1, \ldots, x_n)\}$ in $F[x_1, \ldots, x_n]$, given $F \cup \{x_i\}$ in $O(n \log^2 n)$ total operations and $n \log n$ nonscalar multiplications. The functions $\{p_j\}$ are defined by $p_j(x_1, \ldots, x_n) = \sum_{i_1 < i_2 \cdots < i_j} x_{i_1} x_{i_2} \cdots x_{i_j}$ and satisfy

$$\prod_{i=1}^{n} (t - x_i) = t^n - p_1(\vec{x})t^{n-1} + \cdots + (-1)^n p_n(\vec{x}).$$

We now show that fast evaluation is reducible to fast division (and multiplication).

THEOREM 4.5.3

Let u have precision n and let $\{m_1, \ldots, m_n\}$ be single precision. Then $\{r_i = u \bmod m_i \mid 1 \leq i \leq n\}$ can be computed in $O(D(n) \log n) = O(M(n) \log n) = O(n \log^2 n)$ steps.

Proof. By Theorem 4.5.1 we can assume that all required supermoduli have been precomputed. The idea is to use the division algorithm along the lines suggested by Fiduccia (1972b). Namely, if $u = qv + r$ and $v \bmod m_i = 0$, then $u \bmod m_i = r \bmod m_i$. We choose the divisors

Fast Evaluation and Interpolation

$v_1 = \prod_{i=1}^{n/2} m_i$ and $v_2 = \prod_{n/2+1}^{n} m_i$; then let $u = q_1 v_1 + r_1$ and $u = q_2 v_2 + r_2$ define r_1 and r_2. It follows that

$$u \bmod m_i = \begin{cases} r_1 \bmod m_i & 1 \leq i \leq n/2 \\ r_2 \bmod m_i & n/2 + 1 \leq i \leq n \end{cases}$$

Moreover, we can assume that the precision of each r_i is roughly $n/2$ (for the domain $F[x]$, $\deg r_i \leq \deg v - 1$, and therefore precision $r_i \leq$ precision $v - 1 = n/2$). Since we assume that v_1 and v_2 have been precomputed, we can apply the divide and conquer strategy; the resulting complexity analysis is

$$T(n) = 2T(n/2) + \text{OVERHEAD}(n)$$

$$= 2T(n/2) + 2D(n) = O(D(n) \log n) = O(n \log^2 n).$$

COROLLARY 4.5.4

Let $p(x)$ be any nth degree polynomial in $F[x]$. We can compute $\{p(x_1), p(x_2), \ldots, p(x_m) \mid n \leq m\}$ in $F[x_1, \ldots, x_m]$, given $F \cup \{x_1, \ldots, x_m\}$, in $O(m \log^2 n)$ total operations and $O(m \log n)$ nonscalar multiplications.

The next two theorems show that interpolation can be reduced to evaluation (and multiplication). This reduction was first discovered by Horowitz (1972) for the domain $F[x]$ with the bound $O(M(n) \log^2 n)$.

Consider the Lagrangian interpolation formula in $F[x]$:

$$u(x) = \sum_{k=1}^{n} y_k a_k \prod_{\substack{i \neq k \\ 0 \leq i \leq n}} (x - x_i)$$

where

$$y_k = u(x_k) \quad \text{and} \quad a_k = 1 \Big/ \prod_{\substack{i \neq k \\ 0 \leq i \leq n}} (x_k - x_i).$$

This generalizes to any Euclidean domain (see Lipson [1971] so that

$$u = \sum_{k=1}^{n} y_k a_k \prod_{\substack{i \neq k \\ 1 \leq i \leq n}} m_i,$$

where the m_i are relatively prime, $y_k = u \bmod m_k$, and $a_k = (\prod_{i \neq k} s_{ik}) \bmod m_k$, where $s_{ik} = m_i^{-1} \bmod m_k$ (i.e., $s_{ik} m_i = 1 \bmod m_k$). If the m_i are single precision, then so are the $\{a_k\}$. To see that this is the Lagrangian formula, we have:

LEMMA 4.5.5

(a) $(a_k \prod_{i \neq k} m_i) \bmod m_k = 1$.
(b) $(a_k \prod_{i \neq k} m_i) \bmod m_j = 0$ for $j \neq k$.

THEOREM 4.5.6 (preconditioned interpolation)

Assume that $\{a_k \mid 1 \leq k \leq n\}$ has been precomputed; then u (of precision n) can be computed from the single precision inputs $\{y_k \mid 1 \leq k \leq n\}$ in $O(M(n) \log n)$ steps.

Proof. Again we can assume that all required supermoduli have been precomputed. Now we split the problem as indicated:

$$\sum_{k=1}^{n} y_k a_k \prod_{\substack{i \neq k \\ 1 \leq i \leq n}} m_i$$

$$= \sum_{k=1}^{n/2} y_k a_k \prod m_i + \sum_{k=n/2+1}^{n} y_k a_k \prod m_i$$

$$= \prod_{i=n/2+1}^{n} m_i \sum_{k=1}^{n/2} y_k a_k \prod_{\substack{i \neq k \\ 1 \leq i \leq n/2}} m_i$$

$$+ \prod_{i=1}^{n/2} m_i \sum_{k=n/2+1}^{n} y_k a_k \prod_{\substack{i \neq k \\ n/2+1 \leq i \leq n}} m_i$$

Fast Evaluation and Interpolation

Binary recursion again yields the analysis $T(n) = 2T(n/2) +$ OVERHEAD $(n) = 2T(n/2) + 2M(n/2) = O(n \log^2 n)$.

For the complete interpolation, it remains only to show how to efficiently precompute the $\{a_k\}$.

THEOREM 4.5.7

Given the relatively prime single precision $\{m_k \mid 1 \leq k \leq n\}$, we can compute $\{a_k \mid 1 \leq k \leq n\}$ in $O(M(n) \log n) = O(n \log^2 n)$ steps.

Proof. We can again assume that $M \overset{\circ}{=} \prod_{i=1}^{n} m_i$ has been precomputed. Let $b_k = (\prod_{i \neq k} m_i) \bmod m_k$; by Lemma 4.5.5, we then have $1 = b_k a_k \bmod m_k$. We can compute $\{a_k\}$ from $\{b_k\}$, using the classical extended Euclidean algorithm (see Knuth [1969], p. 302]. Since each m_k and b_k is single precision, such a computation requires only a constant number of operations, and therefore only $O(n)$ operations, to compute all the $\{a_k\}$ from the $\{b_k\}$.

By definition of b_k we have

$$\prod_{i \neq k} m_i = d m_k + b_k,$$

which implies

$$\prod_{i=1}^{n} m_i = M = d m_k^2 + b_k m_k,$$

so that

$$b_k m_k \equiv M \bmod m_k^2.$$

The computation of $\{M \bmod m_k^2 \mid 1 \leq k \leq n\}$ is just the evaluation problem (except here m_k^2 is double precision, but asymptotically this does not affect the complexity). Finally, we can compute the $\{b_k \mid 1 \leq k \leq n\}$ from $\{b_k m_k \mid 1 \leq k \leq n\}$, using exact division in $O(n)$ steps, since each division involves only a double precision dividend and single precision divisor.
□

In the polynomial setting, Horowitz (1972) uses a more direct reduction to evaluation for computing the $\{a_k\}$. With $m_i = (x - x_i)$, we have $a_k = 1/\prod_{c \neq k}(x - x_i) \triangleq 1/b_k$. Given the $\{b_k \mid 1 \leqslant k \leqslant n\}$, we can compute the $\{a_k \mid 1 \leqslant k \leqslant n\}$ in n steps. We have

$$b_k = \prod_{i \neq k}(x - x_i) = \frac{\prod_{i=1}^{n}(x - x_i)}{(x - x_k)} \triangleq \frac{p(x)}{(x - x_k)} = \frac{p(x) - 0}{(x - x_k)}$$

$$= \frac{p(x) - p(x_k)}{(x - x_k)} = p'(x)\big|_{x = x_k}$$

So, to compute $\{b_k \mid 1 \leqslant k \leqslant n\}$ is to evaluate the $n - 1$ degree polynomial $p'(x)$ at n points. This reduction illustrates (as does Section 4.4) how different domains may have different computational properties.

THEOREM 4.5.8

Given the relatively prime $\{m_i \mid 1 \leqslant i \leqslant n\}$ and the $\{y_k \mid 1 \leqslant k \leqslant n\}$, all single precision, we can compute u, satisfying the Chinese remainder theorem, in $O(M(n) \log n) = O(n \log^2 n)$ steps.

Proof. Follows directly from Theorems 4.5.6 and 4.5.7.

COROLLARY 4.5.9

Define the rational functions $c_j \in F(x_1, \ldots, x_n, y_1, \ldots, y_n)$ $0 \leqslant j \leqslant n - 1$ satisfying $\sum_{j=0}^{n-1} c_j x_k^j = y_k$, $1 \leqslant k \leqslant n$. We can compute $\{a_j\}$ in $F(x_1, \ldots, x_n, y_1, \ldots, y_n)$, given $F \cup \{x_i\} \cup \{y_i\}$, using $O(n \log^2 n)$ arithmetic operations and $O(n \log n)$ nonscalar multiplications.

In Chapter 5, we will show that the nonscalar operation bounds in Corollaries 4.5.2, 4.5.4, and 4.5.9 are all asymptotically optimal. It remains an open problem whether or not the extra $\log n$ factors involved in the total operation counts are necessary. However, the analysis in Chapter 5 will not apply to the domain **Z**. The difficulty in trying to establish nonlinear lower bounds relative to a general computational model should be well appreciated. (See Cook and Aanderaa [1969] and the improved version by Paterson et al. [1973]

for a brillant effort concerning the complexity of integer multiplication.) It is possible that a more structured model for multiple precision number theoretic computation would yield meaningful results. It is the algebraic structure implicit in the arithmetic model which makes possible the development presented in Chapter 5.

PROBLEMS

4.1 Describe an algorithm for multiplying nth degree polynomials in $F[x]$, which uses $o(n^2)$ operations and does not rely on any property of the particular field F.

4.2 Let

$$p(x,y) = \sum_{i=0}^{n} \sum_{j=0}^{n} c_{ij} x^i y^j,$$

$$q(x,y) = \sum_{i=0}^{n} \sum_{j=0}^{n} d_{ij} x^i y^j;$$

$p(x,y)$ and $q(x,y)$ in $C[x,y]$. Discuss an asymptotically fast method for computing the coefficients of $w(x,y) \stackrel{\circ}{=} p(x,y) \times q(x,y)$.

Problems 4.3, 4.4, and 4.5 refer to the discussion of fast integer multiplication in Section 4.3.

4.3 Show that $\{w_i \bmod K \mid 0 \leq i \leq 2K - 2\}$ can be computed in $O(K^{\log_2 3})$ steps.

4.4 Show that $\{w_i \mid 0 \leq i \leq 2K - 2\}$ can be computed from $\{w_i \bmod K\}$ and $\{w_i \bmod 2^{2L} + 1\}$ in $O(N)$ steps.

4.5 Let $M = 2^{2L} + 1$, $r = 2K$, $\omega = 2^{4L/r}$, and assume r divides $4L$. Show that in Z_M

(a) $\displaystyle\sum_{0 \leq s < i} \omega^{s\alpha} = \begin{cases} r & \text{if } \alpha = 0 \bmod r; \\ 0 & \text{otherwise} \end{cases}$

(b) r^{-1} exists.

4.6 (Aho et al. [1975].) Suppose the points $\{x_0, \ldots, x_n\}$ can be used for a fast transform (i.e., $\sum_{i=0}^{n} a_i x_j^i$, $1 \leq j \leq n$ can be computed in $O(n \log n)$ operations). Show that for any h, the points $\{x_0 + h, \ldots, x_n + h\}$ can also be used for a fast transform.

4.7 Show that an nth degree polynomial and all its (normalized) derivatives can be computed in $O(n \log n)$ arithmetic operations. (Compare with Section 2.4, but do not infer that the method implied here should be used rather than, say, the Shaw and Traub [1974] algorithm.)

CONJECTURES—OPEN PROBLEMS

1. Show that polynomials (in the usual coefficient representation) cannot be multiplied (or divided) in $O(n)$ arithmetic operations.
2. Can the arithmetic operation bounds in Corollaries 4.5.2, 4.5.4, and 4.5.9 be reduced to $O(n \log n)$?
3. Can an nth degree polynomial and all its derivatives be evaluated in $O(n)$ arithmetic operations?

Chapter 5

NONLINEAR LOWER BOUNDS

The fundamental barrier throughout "concrete complexity" is in proving nonlinear lower bounds on "naturally defined" computational problems. We measure complexity relative to the "size of the input presentation," and we often say that linear lower bounds are obtainable because every piece of the input must be appropriately accessed. Nonlinear lower bounds are said to express something more intrinsic about complexity. For complexity based on the multitape Turing machine, the only nonlinear lower bounds attainable thus far are directly or indirectly based on diagonalization, although there are results that apply to interesting but restricted models (e. g., on-line computation, one-tape machines). For arithmetic complexity we can hope to exploit the algebraic structure of the computational steps.

5.1 Lower Bounds for Nonscalar * Operations

We have already studied a problem in Chapter 3 which can (with some stretch of the imagination) be interpreted as possessing a nonlinear lower bound. Namely, once we have set specific coefficients, a polynomial has only one input parameter, the indeterminate x, and we can exhibit specific nth degree polynomials requiring about $\sqrt{n}$ nonscalar * operations and $(3/2)\,n$ total arithmetic operations. But we can explain this "nonlinearity" by saying that the coefficients represent parameters that have been suppressed but are really needed to express the problem.

What if we consider the computation of x^n? Here we previously argued that a simple growth argument yields a $\log_2 n$ lower bound. Must we somehow dismiss this problem because of its simplicity? Consider an extension of the problem, namely, to compute x^n for $x = x_1, \ldots, x_m$. Is the more obvious approach, raising each x_i to the nth power independently, an optimal (or near optimal) method? Section 4.5 demonstrates that independence proofs do not readily accumulate. What about growth arguments?

Strassen (1973a) shows us that the proper framework for such a question lies within algebraic geometry. What we want is to be able to define "the degree of a set of polynomials." In order to mimic the simple growth argument, we want this notion of degree to satisfy some nice properties; specifically, the degree cannot more than double after a nonscalar * operation, and it does not increase as the result of either a ± operation or a scalar multiplication.

Let us look at a geometric interpretation for degree. A polynomial $p(x) \in F[x]$ can be viewed in terms of its graph $\langle x, p(x) \rangle$ in two-dimensional space F^2. To say that $p(x)$ has large degree is to say that the graph has many "bumps." We can calculate the degree by putting a line through the graph. Under the following conditions, the line will intersect the graph in exactly n points, n = formal degree of $p(x)$ (assuming the leading coefficient is nonzero and $\deg p > 0$):

(a) F is algebraically closed; e.g., $F = \mathbf{C}$.
(b) The line is not tangent to any point on the graph.
(c) The line is not asymptotically parallel to the graph.

The tangent contribution is adjusted for by calculating the multiplicity of the intersection. Alternatively, we can translate the line to achieve a maximum number of points of intersection. To compensate for asymptotes and parallel lines we must extend the graph into two-dimensional projective space where a line at infinity has been added. The important theorem of Bezout generalizes this property to tell us that (in projective space over an algebraically closed field) two curves of degrees n and m will intersect at nm points if we count multiplicities.

To generalize the concept of degree to a set of polynomials or rational functions, we intersect the graph (extended into projective space) by a linear subspace of the appropriate dimension (= codimension of the graph). For example, a surface $\langle x_1, x_2, p(x_1, x_2) \rangle \in F^3$

Lower Bounds for Nonscalar * Operations

is intersected by a line; a curve $\langle x, p_1(x), p_2(x)\rangle \in F^3$ is intersected by a plane. So, then, here is Strassen's approach: Use a generalized version of Bezout's theorem to show that this geometric notion of degree has the desired properties.

Strassen's development relies essentially on some nontrivial concepts and results from algebraic geometry. We will not provide proofs for the necessary lemmas; an understandable introduction to the subject is provided by Fulton (1969) and a good general reference consists of the lecture notes by Shafarevitch (1969). We can, however, appeal to our geometric intuition to see that the mathematics is well motivated. Thinking in terms of C^m (or even R^m) space is usually helpful as long as one realizes that the desired results generally do not hold in that setting. For example, consider the curves

$$C_1 = \{\langle x_1, x_2\rangle | x_1^2 + x_2^2 - 1 = 0\}$$

and

$$C_2 = \{\langle x_1, x_2\rangle | (x_1 - 1)^2 + x_2^2 - 1 = 0\}$$

in (affine) C^2 space. They intersect at the points $\langle 1/2, \sqrt{3}/2\rangle$ and $\langle 1/2, -\sqrt{3}/2\rangle$. The projective curves (see Definition 5.1.1 for notation) $C_1' = \{[x_0, x_1, x_2] | x_1^2 + x_2^2 - x_0^2 = 0\}$ and $C_2' = \{[x_0, x_1, x_2] | (x_1 - x_0)^2 + x_2^2 - x_0^2 = 0\}$ intersect at the four (in accordance with Bezout) points $[1, 1/2, \sqrt{3}/2], [1, 1/2, -\sqrt{3}/2], [0, 1, i], [0, 1, -i]$. Throughout the remainder of this section, F represents an algebraically closed field.

DEFINITION 5.1.1.

A *point in projective* P^N space is an equivalence class of points in $F^{N+1} - \{0, \ldots, 0\}$; the equivalence relation ρ is defined as

$$\langle x_0, \ldots, x_N\rangle \rho \langle y_0, \ldots, y_N\rangle$$

iff $y_i = \lambda x_i$, $0 \leq i \leq N$, $\lambda \in F$, $\lambda \neq 0$.

We will let the zeroth coordinate be the projective coordinate and call $H = \{[z_0, \ldots, z_N] \in P^N | z_0 = 0\}$ the hyperplane at infinity, where $[\langle z_0, \ldots, z_n\rangle]$ is the equivalence class represented by $\langle z_0, \ldots, z_n\rangle$. As an example, consider the parallel projective lines

$L_1 = \{[\langle w_0, w_1, w_2\rangle] \mid w_2 = mw_1 + cw_0\}$ and $L_2 = \{[\langle w_0, w_1, w_2\rangle] \mid w_2 = mw_1 + dw_0\}$. They intersect at the point $[\langle 0, 1, m\rangle] \in H$.

DEFINITION 5.1.2

W is a *closed* (also called *algebraic*) subset of P^N if $W = \{[w] \in P^N \mid p_1(\vec{w}) = p_2(\vec{w}) = \cdots = p_r(\vec{w}) = 0\}$ for some finite set of $N+1$ variable forms (= homogeneous polynomials) $\{p_i \mid p_i \in F[z_0, \ldots, z_N]\}$. As special cases:

(i) If $r = 1$, W is called a *hypersurface*.
(ii) If $r = 1$ and p_1 is a linear form, W is called a *hyperplane*.
(iii) If $p_1, \ldots, p_r$ are all linear forms, W is called a *linear subspace*.

DEFINITION 5.1.3

A closed set W is called *irreducible* if there does not exist a partition $W = W_1 \cup W_2$ with W_1, W_2 closed and $W_1 \neq W \neq W_2$.

A hypersurface $W = \{[\vec{w}] \mid p(\vec{w}) = 0\}$ is irreducible if and only if the form p is equal to p_1^k, p_1 irreducible. Every point $[\vec{w}]$, the common zero of $\{z_0 - w_0, z_1 - w_1, \ldots, z_n - w_n\}$, is irreducible. However, $W = \{[\vec{w}] \mid p_1(\vec{w}) = p_2(\vec{w}) = 0\}$ can be reducible even when p_1 and p_2 are relatively prime irreducible forms (e.g., the common zeros of $p_1 = z_1^2 + z_2 z_0 - 2z_0^2$ and $p_2 = z_1^2 - z_2 z_0$ are the points $[1, 1, 1]$, $[1, -1, 1]$, $[0, 0, 1]$. Indeed almost every two curves will have this property.

LEMMA 5.1.4

Every closed W can be uniquely decomposed into a finite set of irreducible *components* $\{W_i\}$: $W = \cup_{i=1}^{t} W_i$ and $W_i \not\subset W_j$ for all $i \neq j$.

In the case of a hypersurface W determined by the form p, the components will be determined by the unique factorization of $p = \prod_{i=1}^{k} p_i^{n_i}$ (i.e., $W = \cup_i \{[\vec{w}] \mid p_i(\vec{w}) = 0\}$).

Lower Bounds for Nonscalar * Operations

DEFINITION 5.1.5

Let W be closed. The *dimension* $W = \dim W \stackrel{\circ}{=} \max \{ n \mid$ there exist irreducible $\{ C_0, \ldots, C_n : C_0 \subset C_1 \cdots \subset C_n \subseteq W \}$.

This algebraic definition of dimension coincides with our geometric intuition. For example, a two-dimensional surface contains a curve having dimension 1, which in turn contains a point having dimension 0.

LEMMA 5.1.6

Let W be irreducible and let H be an irreducible hypersurface not containing W. Then every component of $W \cap H$ has dimension = $\dim W - 1$.

Our geometric intuition may be somewhat misleading here; e.g., in $\mathbf{R}^3$, a hypersurface and a hyperplane can intersect in a single point. A proof of the lemma and other basic theorems concerning dimension can be found in Shafarevitch (1969, pp. 44-50).

LEMMA 5.1.7

Let W be closed and let all its components have the same dimension n. If $C \subseteq W$ is irreducible and also of dimension n, then C is a component of W.

By the definition of dimension, C cannot be properly contained in any component; if C intersects a component C_1, then irreducibility precludes C from intersecting any components outside of C_1.

DEFINITION 5.1.8

Let $W \subset P^N$ be irreducible. The *degree* $W = \deg W \stackrel{\circ}{=} \max_L$ (cardinality of $L \cap W \mid L$ is a linear subspace, $W \cap L$ finite, $\dim L = N - \dim W$).

LEMMA 5.1.9

Let f be an irreducible form of formal degree m; then the hypersurface determined by $f = \{ \lfloor \vec{w} \rfloor \in P^N \mid f(w) = 0 \}$ has degree m.

We can now see that Definition 5.1.8 corresponds to the usual degree of a multivariate polynomial. For if the formal degree of $p(z_1,\ldots,z_N)$ is m, then the graph p when embedded into projective space P^{N+1} (as in Definition 5.1.10) is determined by the irreducible form $z_0^{m-1} z_{N+1} - \tilde{p}(z_0,\ldots,z_N)$ where $\tilde{p}$ is the *homogenized* form corresponding to p (i.e., if $p = f_{i_1} + \cdots + f_m$, f_i a form of deg j, then $\tilde{p} = f_{i_1} z_0^{m-i_1} + \cdots + f_m$). The graph of the reduced rational function $f = p/q$, $\deg p = m$, $\deg q = n$ is determined either by $z_0^{m-n-1} \times z_{N+1} \tilde{q}(z_0,\ldots,z_N) - \tilde{p}(z_0,\ldots,z_N)$ or by $z_{N+1} \tilde{q}(z_0,\ldots,z_N) - z_0^{n-m+1} \tilde{p}(z_0,\ldots,z_N)$, depending on whether $m \geq n+1$ or $m < n+1$. Then the (algebraic) degree of p/q is max (deg p, deg $q + 1$).

DEFINITION 5.1.10

Let $p_1,\ldots,p_r$ be rational functions in $F(z_1,\ldots,z_n)$.

(a) Graph $(p_1,\ldots,p_r)$
$= \{\langle x_1,\ldots,x_n, p_1(x_1,\ldots,x_n),\ldots,$
$p_r(x_1,\ldots,x_n)\rangle \mid p_1(x),\ldots,p_r(\vec{x})$ are all defined $\} \subseteq F^{n+r}$.

(b) Graph $(p_1,\ldots,p_r)$ is embedded in P^{n+r} via the mapping θ
$: \langle z_1,\ldots,z_{n+r}\rangle \to [\langle 1,z_1,\ldots,z_{n+r}\rangle]$.

(c) $W(p_1,\ldots,p_r) \stackrel{\circ}{=} \overline{\theta \text{ (graph }(p_1,\ldots,p_r))}$,
where $\overline{V}$ represents the closure of $V \subseteq P^N$ relative to the topology induced by Definition 5.1.2.

If $V = \{\vec{x} \mid f(\vec{x}) = 0, f \text{ a polynomial}\} \subseteq F^N$, then $\overline{\theta V} = \{[\vec{z}] \in P^N \mid f^*(\vec{z}) = 0\}$, where f^* is the homogenized form corresponding to f. But if $V = \{\vec{x} \mid f_i(\vec{x}) = 0, 1 \leq i \leq r\}$, it does not necessarily follow that $\overline{\theta V} = \{[\vec{z}] \in P^N \mid f_i^*(\vec{z}) = 0, 1 \leq i \leq r\}$.

LEMMA 5.1.11

(a) $W(p_1,\ldots,p_r) \subseteq P^{n+r}$ is irreducible.
(b) The degree and dimension of $W(p_1,\ldots,p_r)$ do not depend on the ordering of the $\{p_i\}$; dim $W(p_1,\ldots,p_r) = n$.

Lower Bounds for Nonscalar * Operations

Finally, we have introduced enough machinery to formally express the thrust of Strassen's development. But before giving the proof of Theorem 5.1.14, we need two key lemmas, one being the nonelementary theorem of Bezout. Lemma 5.1.12 is the required application of the general statement found in Strassen (1972).

LEMMA 5.1.12 (Bezout)

Let W be irreducible and H an irreducible hypersurface not containing W. Then $\deg W \cdot \deg H \geq \sum_{i=1}^{t} \deg C_i$, where $C_1, \ldots, C_t$ are the components of $W \cap H$.

LEMMA 5.1.13

Let $\Pi : P^N \to P^{N-1}$ via $[\langle z_0, \ldots, z_N \rangle] \mapsto [\langle z_0, \ldots, z_{N-1} \rangle]$. Suppose $V \subseteq P^{N-1}$, $W \subseteq P^N$ are both irreducible. Then

(a) $\deg \Pi^{-1} V = \deg V$; $\dim \Pi^{-1} V = \dim V + 1$.

(b) $\deg \Pi W \leq \deg W$; $\dim \Pi W = \dim W - 1$ or $\dim W$.

This lemma is well motivated from a geometrical point of view.

THEOREM 5.1.14 (Strassen)

The number of nonscalar * operations required to compute $\{p_1, \ldots, p_r\}$ in $F(x_1, \ldots, x_n)$, given $F \cup \{x_i\}$, is $\geq \lceil \log_2 (\deg W(p_1, \ldots, p_r)) \rceil$.

Proof. It suffices to show the following:
  (a) $\deg W(p_1, \ldots, p_r) \leq \deg W(p_1, \ldots, p_r, p_{r+1})$.
  (b) $\deg W(p_1, \ldots, p_r, p, q, p*q) \leq 2 \deg (p_1, \ldots, p_r, p, q)$.
  (c) $\deg W(p_1, \ldots, p_r, p, q, c_1 p + c_2 q) = \deg (p_1, \ldots, p_r, p, q)$
      for any $c_1, c_2 \in F$.

We will prove just (b). (a) follows directly from Lemma 5.1.13 and the proof of (c) is then a direct analogy of (b). Let $N = n + r + 3$ and

define
$$V = W(p_1, \ldots, p_r, p, q) \subset P^{N-1};$$

$$T = W(p_1, \ldots, p_r, p, q, p * q) \subset P^N.$$

If $*$ is $\times$, then let H be the (degree 2) hypersurface defined by the form $p(z_0, \ldots, z_N) = z_0 z_N - z_{N-1} z_{N-2}$. If $*$ is $\div$, then let H be the hypersurface defined by the form $p(z_0, \ldots, z_N) = z_N z_{N-1} - z_{N-2} z_0$.

(i) $T \subseteq H \cap \overline{\Pi^{-1} V}$ by definition.
(ii) T is irreducible and has dimension n by Lemma 5.1.11.
(iii) $\overline{\Pi^{-1} V}$ is irreducible and has dimension $n+1$ by Lemmas 5.1.11 and 5.1.13.
(iv) $H \cap \overline{\Pi^{-1} V}$ has dimension n by Lemma 5.1.6, since $\overline{\Pi^{-1} V} \not\subset H$.
(v) T is a component of $H \cap \overline{\Pi^{-1} V}$ by Lemma 5.1.7.
(vi) $2 \deg V = 2 \deg \overline{\Pi^{-1} V} = \deg H \cdot \deg \overline{\Pi^{-1} V} \geqslant \deg T$ by Lemmas 5.1.13 and 5.1.12.

□

The next lemma (which is also used in the proof of Lemma 5.1.13) provides an operational means by which to establish lower bounds on the degree.

LEMMA 5.1.15

Let L be a linear subspace not containing the irreducible set W. Then $\deg W \geqslant t =$ number of components of $W \cap L$.

The operational aspect of the lemma derives from the following method: Consider graph $(p_1, \ldots, p_r) \subseteq F^{n+r}$ and suppose we can find a linear subspace L (of F^{n+r}) which intersects graph $(p_1, \ldots, p_r)$ in k points. Then when we embed both L and graph $(p_1, \ldots, p_r)$ into P^{n+r}, each of the intersection points becomes a component. Hence, $W(p_1, \ldots, p_r) \geqslant k$.

We are now able to establish some important concrete corollaries, which more than justify this development. In each case, we want to

Lower Bounds for Nonscalar * Operations

intersect graph $(p_1, \ldots, p_r)$ with an appropriate linear subspace of F^{n+r}.

COROLLARY 5.1.16 (see Corollaries 4.5.2, 4.5.4, 4.5.9)

(a) To compute the elementary symmetric functions $p_1, \ldots, p_n$ in n variables $x_1, \ldots, x_n$ in $F(x_1, \ldots, x_n)$, given $F \cup \{x_i\}$, requires at least $[\log n!] \sim n \log(n/e)$ nonscalar * operations. The p_j are defined by

$$p_j(x_1, \ldots, x_n) = \sum_{i_1 < i_2 < \ldots < i_j} x_{i_1} x_{i_2} \cdots x_{i_j}.$$

(b) Let $p(x) \in C[x]$, degree $p = m$. To compute $\{p_i(x_1, \ldots, x_n) | p_i(\vec{x}) = p(x_i)\ 1 \leq i \leq n\}$ in $C(x_1, \ldots, x_n)$, given $C \cup \{x_i\}$, requires $\geq \lceil n \log m \rceil$ nonscalar * operations. In particular, note that the result holds for $p(x) = x^m$, the "simplest" mth degree polynomial.

(c) Interpolation. Let $\{p_i(x_0, \ldots, x_n, y_0, \ldots, y_n) | 0 \leq i \leq n\} \subseteq F(\vec{x}, \vec{y})$ be the rational functions determined by the equations

$$\sum_{j=0}^{n} p_j(\vec{x}, \vec{y}) x_i^j = y_i \quad (0 \leq i \leq n).$$

To compute $\{p_i\}$ in $F(\vec{x}, \vec{y})$ given $F \cup \{x_i\} \cup \{y_i\}$, requires at least $(n+1) \log n$ nonscalar * operations.

Proof. In each case we want to show that the degree of the set of required functions is sufficiently large. We will prove just (b).

(b) Since C has characteristic 0, we can find a $c \in C$ such that $p(x) - c$ has n distinct roots. Let F^{n+m} be coordinatized by $\langle y_1, \ldots, y_{n+m} \rangle$ and let L be the linear subspace of P^{n+m} determined by $y_{n+i} = c$. Then L will intersect graph $(p_1, \ldots, p_n)$ in m^n points; i.e., at the points $\{\langle x_1, \ldots, x_n, p(x_1), \ldots, p(x_n)\rangle \mid x_j$ is a root of '$p(x) - c$'$\}$. □

We can therefore conclude that the nonscalar operation bounds of Corollaries 4.5.2, 4.5.4, and 4.5.9 are all asymptotically optimal (and, in fact, the bound in 4.5.2 is optimal to within a lower order $O(n)$ term). Whether or not $O(n \log^2 n)$ total operations are needed for any of the problems specified in Corollary 5.1.16 remains unresolved. Indeed, we cannot yet prove that any of these problems require $O(n \log n)$ ± operations (or scalar multiplications).

One naturally wonders if this development necessarily requires the depth and sophistication of algebraic geometry. Could we not find a more "syntactic" means by which we could define the degree of a set of polynomials? Our experience suggests that algebraic geometry is necessary. Without Theorem 5.1.14, we do not even know how to show that $\{x_1^n, x_2^n\}$ requires $2 \log n$ * operations (assuming, of course, that ± operations are allowed). Moreover, from a syntactic point of view, it is not at all clear why the functions defined by an nth degree polynomial and all the derivatives (or by polynomial division) have relatively small degree; and this must be the case, since the problems are computable in $O(n)$ multiplications.

Recently, Strassen (1975) has been able to apply his degree argument to a problem involving branching when he shows that the Schonhage (1971) – Moenck (1973) GCD algorithm is "uniformly optimal" for computing the Euclidean representation of a pair of univariate polynomials. Here, we are not just computing one set of functions since even the length of the remainder sequence depends on the given polynomials. Yet, in spite of the elegance and depth of Strassen's approach, those who are hoping for a nonlinear lower bound for matrix multiplication (or for any set of quadratic polynomials) will have to look elsewhere. For the algebraic degree induced by m quadratic forms (in the sense of Definitions 5.1.8 and 5.1.10) is $\leq 2^m$.

5.2 Attempts to Establish Nonlinear Lower Bounds for Additions and Scalar Operations

The results of Section 5.1 still leave open a number of important questions. Besides matrix multiplication, we would like to know whether the fast Fourier transform provides an asymptotically optimal method for polynomial multiplication (and division); moreover,

Establishing Nonlinear Lower Bounds

we would like to have a bound on the complexity of the FFT. For these problems, and the corollaries of Section 4.5, it is the scalar multiplications and ± operations that appear to be the limiting factor. In this section we will first discuss Morgenstern's (1973a) approach, which yields nonlinear lower bounds in a restricted model. We then conclude with a brief discussion of the approach in Borodin and Cook (1974).

Morgenstern is concerned with the computation of sets of linear forms, say in $F[x_1, \ldots, x_n]$. By definition, the FFT is such a computation; i.e., with $F = C$. We can also view polynomial or matrix multiplication as such a computation either by substituting for one set of variables, say $\{a_i\}$, or by viewing the computed functions to be linear in $F[b_1, \ldots]$ with $F = G(a_1, \ldots)$, G the underlying field.

DEFINITION 5.2.1

A program P in $F[x_1, \ldots, x_n]$, given $F \cup \{x_1, \ldots, x_n\}$, is called a *linear program* if every operation is either a ± step or a scalar multiplication (i.e., one operand is an element of F).

By the method of Theorem 2.5.3 we know that linear programs are asymptotically optimal with respect to total operations when computing sets of linear (or linear affine) functions. Morgenstern (1973b) establishes another interesting property of linear programs.

THEOREM 5.2.2 (see Problem 5.2)

If a general arithmetic program P computes a set S of linear functions using k steps and t nonscalar steps (counting all operations except scalar multiplications), then we can construct (effectively from P) a linear program P' which computes S and uses only t nonscalar steps and at most $2k$ total steps.

It is worth noting that the FFT is computed by a linear program with C as the scalar field in the usual setting (i.e., forgetting the FFT over a finite ring or field). Indeed, the scalars used in the FFT are all roots of unity; i.e., all the scalar multiplications have an operand with modulus equal to 1. Morgenstern (1973a) shows that any linear program for the FFT which uses "small" scalars necessarily requires $O(n \log n)$ operations.

THEOREM 5.2.3

Let the linear program P compute a set $S = \{f_1, \ldots, f_r\}$ of linear affine functions in $F[x_1, \ldots, x_n]$, F a normed field (for definiteness, let $F = \mathbf{Q}, \mathbf{R},$ or $\mathbf{C}$). Let S be defined by the matrix $A \in F_{t \times n}$ and $\vec{\tau} \in F_{r \times 1}$, i.e.,

$$\begin{pmatrix} f_1 \\ \vdots \\ f_r \end{pmatrix} = A \begin{pmatrix} x_1 \\ \vdots \\ x_n \end{pmatrix} + \begin{pmatrix} \tau_1 \\ \vdots \\ \tau_r \end{pmatrix}$$

Let $\Delta A \stackrel{\circ}{=} \max(|\det A'|, A'$ is a square submatrix of $A)$. Then P must have at least $\log \Delta A / \log \hat{c}$ operations, where c is the maximum modulus of any scalar in the program P and $\hat{c} = \max(2,c)$. (We assume that all scalars are explicitly introduced and not computed.)

Proof. Let the matrix A_i represent the linear part of the set of linear affine functions computed up to the ith step. $A_0 = I$, the identity matrix, and $A_{i+1} = \begin{pmatrix} A_i \\ s_{i+1} \end{pmatrix}$, where s_j represents the linear part of the function computed on the jth step. By using elementary properties of the determinant, we can show $\Delta A_{i+1} \leq c \, \Delta A_i$ if the $(i+1)$st step is a scalar multiplication and $\Delta A_{i+1} \leq 2\Delta A_i$ if the $(i+1)$st step is a ± operation. For, let A'_j be the square submatrix defining ΔA_j. If s_{i+1} is not used in A'_{i+1}, then

$$\Delta A_{i+1} = \Delta A_i.$$

If

$$A'_{i+1} = \begin{pmatrix} \vdots \\ s_{i+1} \end{pmatrix} = \begin{pmatrix} \vdots \\ \tau s_j \end{pmatrix}$$

Establishing Nonlinear Lower Bounds

then

$$\Delta A_{i+1} = \left| \det \begin{pmatrix} \vdots \\ \tau s_j \end{pmatrix} \right| = |\tau| \left| \det \begin{pmatrix} \vdots \\ s_j \end{pmatrix} \right| \leq c \cdot \Delta A_i.$$

If

$$A'_{i+1} = \begin{pmatrix} \vdots \\ s_j + s_k \end{pmatrix}$$

then

$$\Delta A_{i+1} = \left| \det \begin{pmatrix} \vdots \\ s_j + s_k \end{pmatrix} \right| \leq \left| \det \begin{pmatrix} \vdots \\ s_j \end{pmatrix} \right| + \left| \det \begin{pmatrix} \vdots \\ s_k \end{pmatrix} \right| \leq 2\Delta A_i.$$

If there are k steps in the program, then A must be a submatrix of A_k, and hence $\Delta A \leq \Delta A_k \leq \hat{c}^k \Delta A_0 = \hat{c}^k$; i.e., $k \geq \log \Delta A / \log \hat{c}$. □

COROLLARY 5.2.4 (Morgenstern)

Let P be any linear program computing the FFT over $\mathbf{C}$; i.e., P computes

$$A\vec{x} = \begin{pmatrix} 1 & 1 & 1 & \cdots & 1 \\ 1 & \omega & & & \omega^{n-1} \\ & & & & \\ 1 & \omega^{n-1} & & & \omega^{(n-1)^2} \end{pmatrix} \begin{pmatrix} x_0 \\ \vdots \\ x_{n-1} \end{pmatrix}$$

Suppose the maximum modulus of any scalar used in P is 1 (as in Theorem 4.2.3). Then P has at least $(1/2)n \log n$ steps.

Proof. It suffices to show that $\Delta A = n^{n/2}$. The geometric interpretation of the | determinant | of an n by n matrix (over **R**) is that it represents the volume of the n dimensional parallelepiped generated by the row (or column) vectors (see Birkhoff and MacLane [1953], pp. 283, 307). Over **C**, a similar development can be established. Here the inner product of vectors $\vec{a}$ and $\vec{\beta}$ is defined as $\langle a, \beta \rangle = \sum_{i=1}^{n} a_i \beta_i^*$, where β_i^* is the complex conjugate of β_i. From Lemma 4.2.3, we see that the row vectors of A are orthogonal; i.e.,

$$\sum_{s=0}^{n-1} \omega^{i \cdot s} \cdot (\omega^{j \cdot s})^* = \sum_{s=0}^{n-1} \omega^{is} \cdot \omega^{(n-j)s}$$

$$= \sum_{s=0}^{n-1} \omega^{(n+i-j)s}$$

$$= 0 \quad \text{if } 0 \leq i \neq j < n.$$

So, the "volume" will be the product of the lengths of the row vectors $= \prod_{i=1}^{n} n^{1/2} = n^{n/2}$, noting that $\sum_{i=0}^{n-1} \omega^{is} \cdot (\omega^{is})^* = \sum_{i=0}^{n-1} \omega^{ns} = n^{1/2}$. More formally, the matrix with entries $\omega^{ij} n^{-1/2}$ is unitary, and thus all its eigenvalues have absolute value 1; the determinant (ignoring sign) is the product of the eigenvalues (counting multiplicities). □

The importance of Morgenstern's development should now be apparent. Unfortunately, the dependence of the results on the size of the scalars is quite severe. (Given the scalar 1, we can compute the scalar n in one addition and $\log n$ multiplications, or the scalar 2^n in n multiplications.) How much big scalars can help when computing $A\vec{x}$, with the entries of A small as in the FFT, remains an open question.

We conclude this section (and our discussion of the sequential complexity of arithmetic computations) by suggesting one other approach that might lead to some nonlinear lower bounds for ± operations. The idea can be motivated to some extent by the development in Section 5.1, where Strassen shows that we cannot build up a large "degree" without enough, at least log (degree), ∗ operations. To get anywhere, "degree" had to be defined correctly. We are now looking for an analogous property for ± operations. That is, we want

Establishing Nonlinear Lower Bounds

to say: If a polynomial p (or set of rational functions) has "blah" t, then at least $\log t \pm$ operations are required to compute p. One such property might be "the number of distinct real roots," noting that multiplication of polynomials does not increase the associated set of roots.

Recall the "canonical ± program" given in Lemma 3.2.4:

$$t_0 \leftarrow 1$$
$$s_0 \leftarrow x$$
$$\vdots$$
$$t_i \leftarrow \prod_{j < i} s_j^{m_{j,i}}$$
$$s_i \leftarrow c_i + t_i$$
$$\vdots$$
$$t_{k+1} \leftarrow c_{k+1} \prod_{j \leq k} s_j^{m_{j,k+1}}$$

where $\{c_i\} \subseteq F$ and $\{m_{j,i}\} \subseteq \mathbf{Z}$. Here we are interested in computing $p(x) \in F(x)$ in $F(x)$, given $F \cup \{x\}$. The next theorem applies when $F = \mathbf{R}$; an analogous development for $F = \mathbf{C}$ has not yet been established.

THEOREM 5.2.5 (Borodin and Cook [1974])

Let $p(x) = \sum_{i=1}^{N} a_j s_0^{r_{0,j}} \cdots s_n^{r_{n,j}} q_j(s_0, \ldots, s_m)$, where $\{r_{i,j}\} \subseteq \mathbf{Z}$, $a_j \in \mathbf{R}$ and each $q_j \in \mathbf{R}(y_0, \ldots, y_m)$ with degree $\leq M$. (Degree q can be defined, as in Sections 2.4 and 5.1.) Then the number of real zeros plus the number of real poles of p is less than some $\phi(n, N, M)$. Note $\phi(n, N, M)$, which will be bounded by the inductive proof, is independent of the exponents $r_{i,j}$.

The theorem is applied as follows:

COROLLARY 5.2.6

Let $\rho(k)$ be the maximum number of distinct real roots and poles in any $p(x) \in \mathbf{R}(x)$ computable in $k \pm$ operations. Then $\rho(k) \leq \phi(k, 1, 0)$.

Sketch of proof of theorem. The proof consists of a double induction. The main induction is on n; the second induction is on N. Let q' denote dq/dx. To establish the basis $n = 0$, one uses as a lemma the fact that any polynomial $p(x)$ with k nonzero coefficients has at most $2k - 1$ distinct real zeros. The induction step on n has as its basis the case $N = 1$. Here we expand $s_{n+1} = c_{n+1} + \prod_{s \leq n} s_j^{m_{j,i}}$. For the induction on N, we need two lemmas. First (generalizing Rolle's theorem), if p in $F(x)$ has k distinct real zeros plus poles, then p' has at least $(k - 1)/2$ distinct real zeros plus poles. Second, the fact that $s'_{n+1} = q(s_0, \ldots, s_{n+1})$, where $q \in \mathbf{R}(y_0, \ldots, y_{n+1})$, has a degree bounded by some $\psi(n)$, independent of the $\{m_{j,i}\}$. □

The complexity in the statement of Theorem 5.2.5 may be necessary. For, given any t, we can find polynomials $p_1(x)$ and $p_2(x)$ so that neither has any real roots, but so $p_1(x) + p_2(x)$ has t distinct real roots.

FACT 5.2.7

With ρ defined as in Corollary 5.2.6, we have $\rho(k) \geq 3^k$. (The proof is left as Problem 5.4.)

If we could show that $\rho(k) = 3^k$ (or $\rho(k) \leq c^k$ for some c), we would then be very tempted to extend this property to a set of multivariate polynomials. For example, let $p(x)$ have n distinct real zeros, the multivariate polynomial $\tilde{p}(x_1, \ldots, x_n) = p(x_1)^2 + \ldots + p(x_n)^2$ has n^n distinct real zeros lying on an n-dimensional grid in $\mathbf{R}^n$. We suspect that such a multivariate polynomial, and hence $\{p(x_1), \ldots, p(x_n)\}$, requires $0(n \log n) \pm$ operations for its computation.

Finally, for complex polynomials we suggest the following property (instead of the number of distinct real roots). Let $p(z) \in \mathbf{C}[z]$:

$$d(p) \stackrel{\circ}{=} \min\{r \,|\, \text{there is a } q(x,y) \in \mathbf{R}[x,y] \text{ of degree } r \neq 0$$

which "covers" the zeros of p; i.e., if

$$z = x + iy \text{ and } p(z) = 0, \text{ then } q(x,y) = 0\}.$$

For example, the zeros of $p(z) = z^n - 1$ are covered by $x^2 + y^2 = 1$ (independent of n). We conjecture that $d(p)$ is bounded by the number of $\pm$ operations used to compute P.

PROBLEMS

5.1. (Strassen [1973a].) Let

$$f_j(x_1, \ldots, x_n) = \sum_{i=1}^n x_i^j \quad 1 \leq j \leq n.$$

Establish upper and lower bounds on the nonscalar complexity for computing $\{f_j\}$.

5.2 (Morgenstern [1973b].) By considering the differential of every step, show how to convert a program P in $F(x_1, \ldots, x_n)$, given $F \cup \{x_j\}$ for computing linear functions $\{f_i(x_1, \ldots, x_n)\}$, to a linear program P' for $\{f_i\}$. Show that if P has k_1 scalar (with maximum modulus c) multiplications, k_2 $\pm$ operations and k_3 nonscalar operations, then P' has $(k_2 + k_3)$ $\pm$ operations and $k_1 + 2(k_2 + k_3)$ scalar (with maximum modulus c) multiplications.

5.3. Given $A_{n \times n}$, let $A_{n^2 \times n^2}$ denote

$$\begin{pmatrix} A & & & \\ & A & & \\ & & \ddots & \\ & & & A \end{pmatrix}.$$

Why can't we conclude from Theorem 5.2.3 that a $\{0,1\}$ bilinear program for

$$A_{n \times n} X_{n \times n} = A_{n^2 \times n^2} \begin{pmatrix} x_{11} \\ \vdots \\ x_{1n} \\ \vdots \\ x_{nn} \end{pmatrix}$$

requires $n^2 \log n$ steps by letting $A_{n \times n}$ be the matrix defined by the Fourier transform? (Note: $\Delta \hat{A}$ is $n^{n^2/2}$ in this case.)

5.4 (a) For all k, show that there is a polynomial p in $\mathbf{R}[x]$ having 3^k distinct real roots, which can be computed using $k \pm$ operations.

(b) Let $u(k)$ be the maximum number of distinct real roots in any $p(x)$ in $\mathbf{R}[x]$ computable in $k *$ operations. Show $u(k) = 2^k$.

5.5. Complete the proof of Theorem 5.2.6 and derive an upper bound for $\phi(n,N,M)$.

CONJECTURES–OPEN PROBLEMS

1. Suppose P is a linear program computing the functions defined by $A\vec{x}$, and let $||A||$ be the norm defined by max $\{|a_{ij}|\}$. Must P use at least $c (\log \Delta A/\log||A||) \pm$ operations and scalar multiplications for some fixed constant c (say, $c = 1$)?

2. In a recent paper, "Computational Complexity over Finite Fields," Strassen extends his geometric degree method to analyze the nonscalar $*$ complexity (even if one allows branching on zero and counts such an instruction as a multiplication) of the problems in Corollary 5.1.16 when viewed as *finite functions* over $\mathbf{Z}_p$ (or any finite field). It is shown that

(a) There is an nth-degree polynomial p such that (asymptotically) the optimal complexity for evaluating p at $n + 1$ in $0(n \log p)$.

(b) The (asymptotically) optimal complexity for interpolation (here $p \geq n + 1$) is $0(n \log n)$. What is the complexity of computing the elementary symmetric functions over $\mathbf{Z}_p$? What is the complexity of computing $\{x_i^n \mid 0 \leq i \leq n|\}$ over $\mathbf{Z}_p$?

Chapter 6

PARALLEL PROCESSING IN NUMERIC COMPUTATION

In previous chapters we discussed the complexity of evaluating certain arithmetic functions on a serial processor. We now turn our attention to evaluating these functions on a computer that permits several arithmetic operations to occur simultaneously. As is readily seen in the subsequent sections, this area is still very much in the "clever algorithm" stage. In fact, only two (relatively easy) lower bound results are given. The problem of parallel polynomial evaluation is studied in detail; then general rational functions are discussed. Section 6.4 deals with the solution of matrix problems.

6.1 Basic Limitations of Parallelism in Arithmetic Computation

Our model of computation for parallel processing will be, as noted in Chapter 1, a k (independently programmable) processor random access machine. Each processor at the ith (parallel) step is capable of performing any of the basic arithmetic operations on any pair of "previously known" elements. In particular it is assumed that the result of each computation is to be stored in a new register and that such a term may be used by any of the processors in subsequent steps. The processors are separately programmable, and two or more processors may simultaneously access the same register. The model is, then, the same as that used in previous chapters with the modification that it has k independently programmable processors working simultaneously rather than just one. We will refer to such parallelism as k–*parallelism*, and say that a function is k-computable (in A, given B) *in time t*, if k processors working together can compute the function in t parallel steps. One case of k-parallelism of particular interest is that in which there is no a priori bound on k; we call this *unbounded parallelism*.

Parallelism is of greatest benefit when the steps of a computation are clearly independent. Very often, however, we may be able to reorganize a computation, normally considered a serial process, so as to utilize a surprising degree of parallelism. On the other hand, there are many problems in which a high degree of parallelism, or even 2-parallelism cannot be used "efficiently." For example, consider the problem of raising a number to a power n, using only multiplication. Brauer (1939) has shown that under the serial model, x^n can be found in $\log n + O(\log n/\log \log n)$ multiplications; furthermore, Erdös (1960) has shown that $\log n + (1 - \epsilon) \log n/\log \log n$ is a lower bound for "most" n. A simple growth argument (Theorem 6.1.3) says that $\lceil \log n \rceil$ parallel steps are required. Indeed, this bound can be achieved.

LEMMA 6.1.1

x^n is 2-computable from $\{x\}$ in time $\lceil \log n \rceil$.

Proof. One processor successively computes $x^2, x^4, x^8, \ldots, x^{2^r}$, ..., while the other processor accumulates a product (equivalently, the exponents are being summed) of the appropriate powers of x as they are generated. Each term in the accumulated product corresponds to a 1 in the binary representation of n. □

While it is easy to show that $\lceil \log n \rceil$ steps that involve multiplications are needed to compute x^n in $F[x]$, Kung (1974) observed the following curious fact.

THEOREM 6.1.2

The evaluation of x^n over the complex field (or the algebraic closure of any other field) requires only two parallel steps, in which (complex) multiplication or division is performed. Furthermore, this bound is optimal for $n \geq 3$.

Basic Limitations

Proof The construction is based on the identity

$$\sum_{j=1}^{n} \frac{\omega_j}{x - \omega_j} = \frac{n}{x^n - 1},$$

where $\{\omega_j\}$ denotes the set of the nth roots of unity. We compute $(\omega_j/(x - \omega_j))\, j = 1, \ldots, n$ in one parallel step involving only additions, and one involving divisions. Then $(n/(x^n - 1))$ is found in $\lceil \log n \rceil$ more steps, in which only additions are performed. Using the second division, we find $x^n - 1$ and then add 1.

The "optimality" follows from the observation that x^n ($n \geq 3$) cannot be written as a linear combination of terms of the form $(ax + b)/(cx + d)$ and $ex^2 + fx + g$. □

With this bit of counter intuition, it seems the appropriate time to derive some lower bounds. Recall that any rational function $f(x)$ can be written as the ratio of two relatively prime polynomials:

$$f(x) = p(x)/q(x),$$

and that the *degree* of $f(x)$ is defined as max(degree $p(x)$, 1 + degree $q(x)$).

THEOREM 6.1.3 (Kung [1974].)

The computation (in $F(x)$, given $F \cup \{x\}$) of a rational function of degree n, requires at least $\lceil \log n \rceil$ parallel steps.

Proof. The proof proceeds by induction on n. It is trivially true for $n = 1$. Assume that the theorem holds for $n \leq 2^k$. Then, in k steps, only functions of the form $p(x)/q(x)$, where $\max(\overline{\deg p, \deg q + 1})$ $\leq 2^k$, can be computed. Any function computed at the $(k + 1)$st step using an addition or subtraction will be of the form

$$\frac{p_1(x)}{q_1(x)} \pm \frac{p_2(x)}{q_2(x)}$$

$$= \frac{p_1(x)q_2(x) \pm p_2(x)q_1(x)}{q_1(x)q_2(x)}$$

and so is of degree at most 2^{k+1}. A function computed using a multiplication or division at step $k + 1$ is of the form

$$\frac{p_1(x) \cdot p_2(x)}{q_1(x) \cdot q_2(x)} \quad \text{or} \quad \frac{p_1(x) \cdot q_2(x)}{p_2(x) \cdot q_1(x)}$$

and again of degree at most 2^{k+1}

From Lemma 6.1.1 and Theorem 6.1.3 follows. □

COROLLARY 6.1.4

The evaluation of x^n requires $\lceil \log n \rceil$ steps even under unbounded parallelism. This bound can be achieved under 2-parallelism.

COROLLARY 6.1.5

$\{x, x^2, \ldots, x^n\}$ is $\lfloor n/2 \rfloor$-computable in time $\lceil \log n \rceil$ with only $n - 1$ operations being performed. The time and number of operations stated are optimal.

Proof. The algorithm is straightforward; at time t compute $x^{2^{t-1}+1}, x^{2^{t-1}+2}, \ldots, x^{2^t}$. The operation count is minimized, since $n - 1$ different functions are to be computed (x^1 is already known). The time lower bound follows from the theorem. □

Another corollary, perhaps more interesting, involves the computation of (nonlinear) iterations. Consider the simple recurrence relation $y_i = f(y_{i-1})$, where $f(y)$ is a rational function and $y_0 = x$. We observe that if $f(y)$ is linear (i.e., $f(y) = ay + b$), then y_n may be computed in $F(a, b, x)$, given $F \cup \{a, b, x\}$:

Basic Limitations

$$y_n = a(\cdots a(ax+b) + b \cdots) + b$$

$$= a^n x + b \sum_{i=0}^{n-1} a^i$$

$$= a^n x + (b(a^n - 1)/(a-1))$$

$$= a^n(x + b/(a-1)) - (b/(a-1))$$

in $\lceil \log n \rceil + 2$ steps. If f is a linear function of $y_{i-1}, \ldots, y_{i-m}$, the computation is somewhat more complex, but still $O(\log n)$. (See Kogge [1972] and Kogge and Stone [1972] for a more detailed discussion.)

For any rational function $r(x) = p(x)/q(x)$, where $p(x)$ and $q(x)$ are relatively prime polynomials, let us define a slightly different notion of the degree of r: DEG $r(x) = \max(\deg p(x), \deg q(x))$. Then we have:

COROLLARY 6.1.6 (Kung [1974].)

Let $f(y)$ be an arbitrary rational function and $d = \text{DEG } f(y)$. Then, given the recurrence relation $y_i = f(y_{i-1})$ and the initial condition $y_0 = x$, at least $\lceil n \log d \rceil$ parallel steps are required to compute y_n.

Proof. It can be shown that DEG y_n as a function of x is d^n, implying DEG y_n is d^n or $d^n + 1$.

□

A specific consequence of this result is that very little advantage can be taken of parallelism in performing Newton interation for square root $(y_i = 1/2(y_{i-1} + a/y_{i-1}))$. See Winograd (1972), Kung and Traub (1974) for a further discussion of such problems.)

Theorem 6.1.3 is based on a simple growth argument. The following result provides a lower bound based on the serial processing time and a "fan-in argument."

THEOREM 6.1.7 (Munro and Paterson [1973].)

Suppose the computation of a single element Q requires $q \geqslant 1$ binary arithmetic operations. Then the shortest k computation of Q is at least

$\lceil (q + 1 - 2^{\lceil \log k \rceil})/k \rceil + \lceil \log k \rceil$ steps

if $q \geq 2^{\lceil \log k \rceil}$;

and $\lceil \log (q + 1) \rceil$ otherwise.

Proof. Let t be the computation time for some algorithm that computes Q. At time t, at most one processor is usefully employed. At time $t - 1$, at most two processors are usefully employed. In general for all $r \geq 0$, at time $t - r$, at most $\min(k, 2^r)$ processors can compute values to be used in the computation of Q. If

$$q > 2^{\lceil \log k \rceil} - 1 = \sum_{i=1}^{\lceil \log k \rceil} 2^{i-1}$$

(i.e., the boundedness of the parallelism affects the computation), then $t > \lceil \log k \rceil$, and so the total number of useful operations that the algorithm can perform is

$$\underbrace{1 + 2 + 4 + \cdots 2^{\lfloor \log k \rfloor} + k + \cdots k}_{(t \text{ terms})} = k(t - \lceil \log k \rceil) + 2^{\lceil \log k \rceil} - 1.$$

Hence,

$$q \leq k(t - \lceil \log k \rceil) + 2^{\lceil \log k \rceil} - 1,$$

which implies $t \geq \lceil (q + 1 - 2^{\lceil \log k \rceil})/k \rceil + \lceil \log k \rceil$.

On the other hand, if the boundedness of the parallelism does not interfere with the computation, i.e., $t \leq \lceil \log k \rceil$ or $q \leq 2^{\lceil \log k \rceil} - 1$, the maximum number of useful operations that can be performed in t steps is $\sum_{i=0}^{t-1} 2^i = 2^t - 1$. Thus, $q \leq 2^t - 1$ or $t \geq \lceil \log (q + 1) \rceil$.

COROLLARY 6.1.8   □

Under a model of unbounded parallelism, the product of two n-vectors may be found in $\lceil \log n \rceil + 1$ steps; furthermore, this is optimal.

Basic Limitations

Proof. The algorithm is the obvious one. Perform all n multiplications in the first step; then add the n products in $\lceil \log n \rceil$ steps by "fanning in" (i.e., form $a_{2i} + a_{2i+1}$, $i = 1, \ldots, \lfloor n/2 \rfloor$ in the first step, etc.). The optimality follows from the theorem and Winograd's (1970a) result that the product of two vectors requires n multiplications and $n - 1$ additions (e.g., see Theorems 2.3.2 and Corollary 2.2.6).

COROLLARY 6.1.9

The product of two n by n matrices can be performed in $\lceil \log n \rceil + 1$ steps under unbounded parallelism; furthermore, this is optimal.

Proof. Follows from Corollary 6.1.8.

□

When the parallelism is not unbounded, the obvious technique for vector multiplication is still optimal.

COROLLARY 6.1.10

Under k-parallelism the product of two n vectors may be found in $\lfloor (2n - 2^{\lceil \log k \rceil})/k \rfloor + \lceil \log k \rceil$ steps; furthermore, this is optimal for $k < n$.

The problem of matrix multiplication under bounded parallelism is not so easily solved. The technique suggested by the last corollary may be employed, but it is not optimal if k is small, as Strassen's algorithm using only one processor will ultimately be faster unless $k \geq O(n^{3-\log 7}) \simeq O(n^{\cdot 2})$.

The techniques we have suggested thus far for parallel computation have all been relatively straightforward. In Section 6.2 we will present several algorithms for a specific problem, that of the parallel computation of univariate polynomials. We include this material not because of any great importance of the evaluation of high-degree polynomials, but because the approaches taken in some of these algorithms may be useful in more general parallel computations.

6.2 Polynomial Evaluation under Parallelism

Suppose we are to compute $p(x) = \sum_{i=0}^{n} a_i x^i$ in $F(a, x)$, given $F \cup \{x\} \cup \{a_i\}$ under k-parallelism. The standard polynomial evaluation technique, Horner's rule, is very unsuitable for parallel implementation. Indeed, it is seen that at every step the immediately preceding subresult is required. Estrin (1960) and Dorn (1962), whose work is summarized by Knuth (1969), have described techniques similar in nature to Horner's rule which require only about $2\lceil \log n \rceil$ steps. Estrin's algorithm for the evaluation of $p(x)$, an arbitrary nth degree polynomial, recursively computes

$$p(x) = q(x) x^{\lfloor n/2 \rfloor + 1} + r(x),$$

where

$$q(x) = \sum_{i=0}^{\lceil n/2 - 1 \rceil} a_{i + \lfloor n/2 \rfloor + 1} x^i \text{ and } r(x) = \sum_{i=0}^{\lfloor n/2 \rfloor} a_i x^i.$$

This technique has been referred to as *binary splitting* (see Section 4.1) and requires time $2 \log n$ under unbounded parallelism. Dorn describes an algorithm that splits the polynomial into k polynomials and evaluates each by Horner's rule. Let $m = \lfloor (n-i)/k \rfloor$ and

$$q_i(x) = \sum_{j=0}^{m} a_{i + kj} x^{kj} \quad i = 0, \ldots, k - 1.$$

Then

$$p(x) = \sum_{i=0}^{k-1} q_i(x) x^i.$$

With k processors, this algorithm may be implemented in $2n/k + 2 \log k$ steps. Dorn, however, also points out that a major practical problem is in accessing the required information from memory. The assumption of easy access of many words of memory for use in the various processors is implicit in our model.

Polynomial Evaluation

It is natural to wonder whether or not an nth degree polynomial can be evaluated in fewer than $2 \log n$ steps under full parallelism. Both algorithms given have required roughly that time. Estrin's technique requires that the polynomial be split into two subexpressions of essentially equal size. We note, however, that if $q(x)$ and $r(x)$ are formed at the same step, then nothing happens to $r(x)$ while $q(x) x^{\lfloor n/2 \rfloor + 1}$ is being formed. This suggests that $r(x)$ should perhaps be of higher degree (and take one more step to compute) than $q(x)$. Indeed, the golden (Fibonacci) ratio provides a more efficient splitting than does halving (Munro and Paterson [1973] and Muraoka [1971]). By this technique a polynomial of degree less than F_{t+1} may be evaluated in time t. (F_i is the ith member in the Fibonacci sequence 1, 1, 2, 3, 5, 8, 13, 21, ...) Thus, if $T_k(n)$ is the time required to evaluate an nth degree polynomial under k-parallelism,

$$T_\infty (n) \leqslant \alpha \log n + 0(\log n),$$

where

$$\alpha = 1/\log (1/2(\sqrt{5} + 1)) \simeq 1.44.$$

This algorithm is not optimal, however, for large n. Indeed α may be reduced to 1. (See Munro and Paterson [1973] or Maruyama [1973].) The technique in Algorithm 6.2.2 was also used by Brent (1970) for the addition of binary numbers. It may be used as well to compute functions of the form

$$(\ldots (a_n x_n + a_{n-1}) x_{n-1} \ldots a_1) x_1 + a_0$$

in the same time as for $\sum_{i=0}^{n} a_i x^i$.

THEOREM 6.2.1

$$\lceil \log (n+1) \rceil + 1 \leqslant T_\infty(n) \leqslant \log n + 0((\log n)^{1/2}).$$

Proof. The lower bound is a corollary of Theorem 6.1.3, based on the fact that at least $2n + 1$ operations are required if Horner's rule is not used. The upper bound is attained by the following algorithm.

ALGORITHM 6.2.2

Let $D_r = \frac{1}{2}r(r+1) + 1$. We define a recursive evaluation procedure for polynomials of degree n. Let $p = \lceil \log(n+1) \rceil$ and suppose $D_{r-1} < p \leq D_r$. The polynomial may be expressed in the form

$$p(x) = q_0(x) + q_1(x)x^{2^{p-r}} + q_2(x)x^{2 \times 2^{p-r}}$$
$$+ \ldots + q_{2^r-1}(x)x^{(2^r-1)2^{p-r}}$$

where the $q_i(x)$ are polynomials of degree $< 2^{p-r}$. So, to compute $p(x)$, we compute $\{q_i\}$, multiply by the appropriate power of x at the next step and then use r further steps to combine the 2^r terms by binary addition.

To prove inductively that the algorithm takes at most $p + r + 1$ steps, suppose the result is true for all degrees less than n; then $\{q_i\}$ can be computed in time $(p-r) + (r-1) + 1 = p$, since $p - r \leq D_r - r = D_{r-1}$. The powers of x required are also available in this time, so the algorithm runs in time $p + r + 1$.

For the basis of the induction ($n = 1$), we observe that $r = 0$, $p = 1$, and that a polynomial of degree 1 can be computed in two time steps. Since $(r(r-1))/2 < p$, we have shown that

$$T_\infty(n) \leq \log n + (2 \log n)^{1/2} + 0(1).$$

□

This algorithm may be improved somewhat at the cost of further complication; for example, Fig. 6.2.1 illustrates the evaluation of polynomials of degree 21. For polynomials of lower degree, the Fibonacci splitting is in some cases better. Table 6.2.1 compares three methods for programs having at most 16 parallel steps.

TABLE 6.2.1.

Greatest degree of general polynomial computable in time $T_\infty(n) = t$.

$t =$	2	3	4	5	6	7	8	9	10	11	12	13	14	15	16
Fibonacci	1	2	4	7	12	20	33	54	88	143	232	376	609	986	1590
Algorithm 6.2.2	1	1	3	3	7	15	15	31	63	127	127	255	511	1023	2047
Best known	1	2	4	7	12	21	37	63	107	187	327	578	1010	1764	3124

Polynomial Evaluation

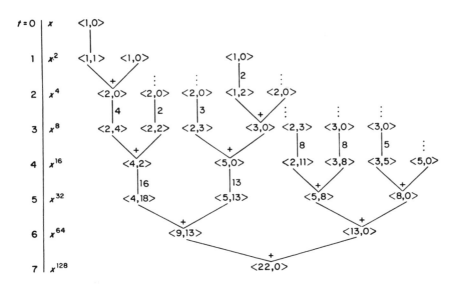

NOTES: $\langle m,n \rangle$ denotes an arbitrary polynomial of the form $c_{m-1}x^{m+n} + \cdots + c_0 x^n$.
$|n$ denotes a multiplication by x^n
$\vdots$ means it has been done similarly elsewhere in the diagram.

Illustrating $T_\infty(21) \leq 7$

After studying the evaluation of polynomials of high degree under unbounded parallelism, we ask what algorithms are fastest when k is small with respect to n. As one would expect, when k is reasonably small, processor utilization can be kept at a rather high level. A bit of thought provides us with optimal algorithms for $k = 2$ or 3. Under these circumstances, it would seem that the evaluation of powers such as x^n is rather wasteful, while the evaluation of x^2 or x^3 is very useful. Thinking along these lines leads to the following algorithm, which is within one step of being optimal for $k \leq n/\log n$.

ALGORITHM 6.2.3 (Munro and Paterson [1973].)

(1) Compute $p_i = a_{2i} + a_{2i+1} x$ for $i = 0, \ldots, \lfloor n/2 \rfloor$ and x^2, $x^4, \ldots, x^{2k-2}, x^{2k}$.

(2) Using processor j ($j = 1, \ldots, k$), compute

$$p_j^* = \sum_{i=0}^{\lfloor n/2k \rfloor} p_{ik+j-1}(x^{2k})^i$$

by Horner's rule.

(3) Finally, evaluate

$$p(x) = \sum_{j=1}^{k} p_j^* x^{2(j-1)}$$

by forming the $p_j^* x^{2(j-1)}$ and then adding them by "fanning in." Note that unless $2k$ exactly divides $n+1$, some of the terms mentioned above vanish, with a consequent saving in arithmetic operations.

To determine the number of steps required by this algorithm, first observe that all processors may be fully utilized until the final fan-in, provided the computing of powers of x can be interlaced with the p_i (i.e., $n \geqslant k \log k$). Hence, the number of steps used in this case is basically that obtained as the optimum in Theorem 6.1.7. An inspection of the algorithm show that n additions and $n+k$ multiplications, or a total of $2n+k$ operations are performed. Therefore, $\lceil ((2n+k+1) - 2^{\lceil \log k \rceil})/k \rceil + \lceil \log k \rceil$ steps are used, provided $n \geqslant k \lceil \log k \rceil$.

Summarizing the preceding algorithm and Theorem 6.1.7, we have shown the following:

COROLLARY 6.2.4

$\lceil ((2n+2) - 2^{\lceil \log k \rceil})/k \rceil + \lceil \log k \rceil \leqslant T_k(n) \leqslant \lceil ((2n+k+1) - 2^{\lceil \log n \rceil})/k \rceil + \lceil \log k \rceil$, provided $n \geqslant k \lceil \log k \rceil$.

We close our discussion of parallel polynomial evaluation with a result found in Munro and Paterson (1973) on "parallel evaluation with preconditioning." Suppose the coefficients of $p(x) = \sum_{i=0}^{n} a_i x^i$ are algebraically independent, and let $T_k^*(n)$ denote the minimum number of steps in which $p(x)$ is k-computable.

Evaluation of Arithmetic Expressions

THEOREM 6.2.5

$T_k^*(n) = 3n/2k + \log k + O(1)$ if $n \geq k$;
$T_\infty^*(n) = \log n + O(1)$.

Proof. That

$$T_k^*(n) \geq \lceil (n + \lceil (n+1)/2 \rceil + 1 - 2^{\lceil \log k \rceil})/k \rceil + \lceil \log k \rceil$$

$$\geq \lceil 3n/2k \rceil + \lceil \log k \rceil - 1,$$

if $n \geq k$, follows from Theorems 3.3.5 and 6.1.7. The computation may be performed in $\lceil 3n/2k \rceil + \lceil \log k \rceil + 1$ steps by factoring the polynomial into k polynomials, each of degree $\leq 2\lceil n/2k \rceil$. By assigning a separate processor to each of these they can be computed in $3\lceil n/2k \rceil + 1$ steps, using the preconditioning technique of Section 3.1. These k terms can be multiplied together in a further $\lceil \log k \rceil$ steps.

That $T_\infty^*(n) \geq \lceil \log n \rceil$ follows from Theorem 6.1.2, while the upper bound of $\lceil \log n \rceil + 3$ is achieved by factoring the polynomial into quadratic terms, computing them, and fanning in with multiplication.

□

6.3 Evaluation of General Arithmetic Expressions

Suppose we are given a general arithmetic expression containing n terms (and so $n - 1$ arithmetic operands) with arbitrary parenthesization. How many steps with unbounded parallelism are necessary to perform the computation? Without essential loss of generality, let us restrict our attention to expressions in which no indeterminate occurs more than once. Our main result is the development of a technique, due to Brent (1974), for evaluation of such expressions in roughly $4 \log n$ steps. Techniques for computing this general type of expression may be applied to the more specific cases in which indeterminates are repeated. For instance, a technique for computing $(a_1 \times a_2 + b_1)/(b_2 \times b_3)$ is readily applied to the evaluation of $(a^2 + b)/b^2$.

Let E denote an expression of the type described and $|E|$ the number of indeterminates in E. We will obtain our results by considering parse trees of such expressions. For instance, if $E = (a_1 \times a_2 + b_1)/(b_2 \times b_3)$, then a parse tree T is given by

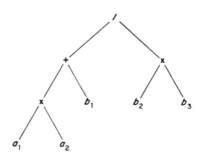

In general, our distinction between an expression and its parse tree will be rather loose. $|T|$ will denote the number of leaves (indeterminates) in a parse tree, and $T = L \wedge R$ will denote that L is the left subtree of T, and R the right subtree. By a subtree rooted at a particular node, we mean the tree rooted at that node and containing *all* of its descendants.

Consider the following straightforward properties of binary trees. Given a binary tree with n leaves, and any integer m, $1 < m \leq n$, we can start at the root of the tree and work down through it until we find a node that is the root of subtree $T' = L \wedge R$ such that $|T'| \geq m$, but $|L| < m$ and $|R| < m$. Similarly, starting at an arbitrary leaf and working up the tree, we will pass nodes that are roots of subtrees of less than m leaves, but then come to a node that is the root of a subtree with at least m nodes. More formally:

LEMMA 6.3.1

If T is a binary tree and $|T| = n$, then for $1 < m \leq n$:
(a) There is a node of T that is the root of the subtree $T' = L_1 \wedge R_1$ such that $|T'| \geq m$, $|L_1| < m$, and $|R_1| < m$.
(b) Every leaf y of T has an ancestor, which is the root of the subtree $T'' = L_2 \wedge R_2$ such that $|T''| \geq m$, but if y is in L_2, $|L_2| < m$ and if y is in R_2, $|R_2| < m$.

Using this lemma we can show

The Complexity of Algebraic and Numeric Problems

THEOREM 6.3.2 (Brent [1974].)
Let

$$t(n) = \begin{cases} n+1 & \text{if } n \leq 2 \\ \lceil 4 \log(n-1) \rceil & \text{if } n \geq 3 \end{cases}$$

Then, if E is a rational expression containing n indeterminates, each of which occurs only once, E may be written as
  (a) $E = F/G$, where F and G are expressions that may be simultaneously evaluated in $t(n) - 2$ parallel steps; and
  (b) $E = (Ay + B)/(Cy + D)$, where y is any indeterminate appearing in E, and A, B, C, D do not depend on y but can be evaluated in $t(n)$ parallel steps.

Proof. We note by inspection that the theorem holds for $n \leq 4$. Suppose the result holds for $n < N$; we now verify that it is true for $n = N$, to prove the theorem by induction.

With $m = \lceil (n+1)/2 \rceil$, apply Lemma 6.3.1(a) to a parse tree for E. We obtain a node x and the subexpression of E, $E(x) = L_1 \theta R_1$:

$$\theta \in \{+, -, \times, \div\}$$

$|E(x)| \geq (n+1)/2$, $|L_1| \leq n/2$, and $|R_1| \leq n/2$.

Let E' be the expression obtained from E by replacing $E(x)$ with a single indeterminate. From the definition of $t(n)$ we note $n \leq 2^{t(n)/4} + 1$; hence,

$$|E'| = n + 1 - |E(x)| \leq (n+1)/2 \leq 2^{(t(n)-4)/4} + 1,$$

and so

$$\lceil 4 \log(|E'| - 1) \rceil \leq t(n) - 4.$$

Applying part (b) of the induction hypothesis to E' gives

$$E = (A_1 E(x) + B_1)/(C_1 E(x) + D_1),$$

where A_1, B_1, C_1 and D_1 can be found in $t(n) - 4$ steps.

We will now apply the first half of the induction hypothesis to L_1 and R_1 to obtain a fast evaluation of $E(x)$. Note that $|L_1|$ and $|R_1|$ are both $\leq n/2 \leq 2^{(t(n)-4)/4} + 1$. By part (a) of the inductive hypothesis, $L_1 = F_1/G_1$ and $R_1 = F_2/G_2$, where F_1, G_1, F_2, G_2 can be found in $t(n/2) - 2 = t(n) - 6$ steps. If we write

$$E(x) = L_1 \theta R_1 = (F_1/G_1)\theta(F_2/G_2) = F_3/G_3,$$

then for

$$\left.\begin{array}{lll} \theta = \text{``}+\text{''} & F_3 = F_1 G_2 + F_2 G_1 \\ \theta = \text{``}-\text{''} & F_3 = F_1 G_2 - F_2 G_1 \\ \theta = \text{``}\times\text{''} & F_3 = F_1 F_2 \end{array}\right\} \text{ and } G_3 = G_1 G_2$$

$$\theta = \text{``}\div\text{''} \quad F_3 = F_1 G_2 \quad \text{and } G_3 = G_1 F_2$$

which implies F_3 and G_3 can be found in $t(n) - 4$ steps. Therefore, as $E = F/G$, $F = A_1 E(x) + B_1$, $G = C_1 E(x) + D_1$ can be found in $t(n) - 2$ steps, completing the proof part (a) of the induction step.

To verify the second part of the induction step, we first apply Lemma 6.3.1(b) for some leaf y, with $m = \lceil(n+1)/2\rceil$. Hence, we have a subexpression of E, $E(z) = L_2 \theta R_2$, such that $|E(z)| \geq (n+1)/2$ and either y is in L_2 and $|L_2| \leq n/2$, or y is in R_2 and $|R_2| \leq n/2$. Without loss of generality, assume y is in L_2.

Let E'' denote the expression obtained from E by replacing $E(z)$ by a single indeterminate. Then $|E''| = n + 1 - |E(z)| \leq 2^{(t(n)-4)/4} + 1$. Applying part (b) of the induction hypothesis to E'' gives

$$E = (A_2 E(z) + B_2)/(C_2 E(z) + D_2),$$

where A_2, B_2, C_2, D_2 can be found in $t(n) - 4$ steps. Similarly, $L_2 = (A_3 y + B_3)/(C_3 y + D_3)$, where A_3, B_3, C_3, D_3 can be found in $t(n) - 4$ steps. Since $|R_2| \leq n - 1$, by the part (a) of the induction hypothesis, $R_2 = F_4/G_4$, and F_4 and G_4, may be evaluated in $t(n) - 2$ steps.

Now, since $E(z) = L_2 \theta R_2$, the expressions given for E, L_2, and R_2 imply we can write $E = (Ay + B)/(Cy + D)$, where

$$A = \begin{cases} (A_2 A_3 + B_2 C_3)G_4 + (A_2 C_3)F_4 & \text{if } \theta = ''+''\\ (A_2 A_3 + B_2 C_3)G_4 - (A_2 C_3)F_4 & \text{if } \theta = ''-''\\ (A_2 A_3)F_4 + (B_2 C_3)G_4 & \text{if } \theta = ''\times''\\ (A_2 A_3)G_4 + (B_2 C_3)F_4 & \text{if } \theta = ''\div'' \end{cases}$$

Similar expressions are obtained for B, C, and D. From this it follows that A (B, C, and D) can be found in $t(n)$ steps.

□

We can rephrase this result in terms of the number of operations appearing in an arithmetic expression.

COROLLARY 6.3.3

An arithmetic expression containing $q \geq 2$ operations may be evaluated in $\lceil 4 \log q \rceil - 1$ parallel steps.

THEOREM 6.3.4

Using the technique described in the proof of Theorem 6.3.2, F and G may be computed using $3(n-1)$ processors and $10(n-1)$ arithmetic operations. Furthermore, A, B, C, and D may be computed with $3n-8$ processors and $10n-29$ arithmetic operations ($n \geq 3$).

The proof follows by induction in exactly the same way as does Theorem 6.3.2. The following interesting lemma is also due to Brent (1974).

LEMMA 6.3.5

If $E_1, \ldots, E_r$ can be computed simultaneously in t parallel steps and q basic operations with unbounded parallelism, then $\{E_i\}$ can be evaluated in time $t + (q-t)/k$ steps, using k processors.

Proof. Suppose s_i operations are performed at time step i ($i = 1, \ldots, t$), so $\sum_{i=1}^{t} s_i = q$. With k processors, we can perform step i of the fully parallel technique in $\lceil s_i/k \rceil$ steps. Therefore, the entire computation can be performed in time $\sum_{i=1}^{t} \lceil s_i/k \rceil \leq (1 - 1/k)t + (\sum_{i=1}^{t} s_i)/k = t + (q-t)/k$.

From this lemma, Theorem 6.3.2, and Theorem 6.3.4, we have: □

COROLLARY 6.3.6

An arithmetic expression containing $q \geq 2$ operations can be evaluated in $4 \log q + 10(q-2)/k$ steps, using k processors.

An interesting observation about the algorithm developed in this section is that only one division is performed, and it is the last step. Therefore, if division is "more costly" than the other operations, this algorithm is particularly useful.

Theorem 6.1.7 (fan-in theorem) provides a lower bound of $\log n$ steps in the fully parallel case and of roughly $n/k - 1 + \log k$ under k-parallelism. Brent et al., (1973) have shown that roughly $2.465 \log n$ steps will suffice in the fully parallel case when divisions are not permitted. Under k-parallelism, Brent (1973) shows $2(n-1)/k + 4 \log n$ steps and, more recently, Winograd (1974) shows $3n/2k + O(\log^2 n)$ steps are sufficient for expressions without division. This latter bound is nearly optimal as Hyafil and Kung (1974) have shown that, ignoring second-order terms, $3n/2(k+1)$ steps are required for all k.

6.4 Parallel Solution of Matrix Problems

In the previous three sections we have described techniques for computing polynomials and "general arithmetic expressions" in a "near optimal" number of parallel steps. Functions of these types are frequently evaluated in practice. Their degrees, however, are usually fairly low; so, some of our algorithms are admittedly esoteric. One class of problems that does occur in practice with "fairly large n" is that of matrix problems. In this class are included the evaluation of matrix products, inverses, and determinants and the solution of systems of linear equations.

In Chapter 2 it was shown that under the serial model of computation, the problems of matrix multiplication and inversion are of the same order of difficulty. In the case of unbounded parallelism, this

Parallel Solution of Matrix Problems

does not appear to be true. Corollary 6.1.9 states that $\lceil \log n \rceil + 1$ parallel steps are necessary and sufficient to multiply a pair of n by n matrices. This section is devoted to the solution of three closely related problems: matrix inversion, determinant computation, and the solution of systems of linear equations.

THEOREM 6.4.1 (Csanky [1974])

With unbounded parallelism, the number of steps required to invert a full $n \times n$ matrix ($I_\infty(n)$), the number to solve n linear equations in n unknowns ($E_\infty(n)$), and the number to compute the determinant of a full $n \times n$ matrix ($D_\infty(n)$), are all of the same order of magnitude.

Proof. First observe that $\det A$, each element of A^{-1}, and each x_i in the solution of $A\vec{x} = \vec{b}$, will generally depend on each of the n^2 elements of the matrix A. It follows from Theorem 6.1.7 that $\lceil 2 \log n \rceil$ is a lower bound on $I_\infty(n)$, $E_\infty(n)$, and $D_\infty(n)$.

We will prove the theorem by showing

(1) $E_\infty(n) \leq I_\infty(n) + \lceil \log n \rceil + 1$

Solve $A\vec{x} = \vec{b}$ by computing A^{-1} and then using $\lceil \log n \rceil + 1$ more steps to form $\vec{x} = A^{-1}\vec{b}$.

(2) $I_\infty(n) \leq D_\infty(n) + 1$

Compute A^{-1} as adjoint $A/\det A$. Since each element of the adjoint is the determinant of an $(n-1)$ by $(n-1)$ matrix, the entire process takes $D_\infty(n) + 1$ steps.

(3) $D_\infty(n) \leq E_\infty(n) + \lceil \log n \rceil$

Let $\tilde{A}_i$ denote the i by i submatrix of A containing elements $\{ a_{j,k} | j, k \geq n+1-i \}$.

Hence $\tilde{A}_n = A$ and $\tilde{A}_i = a_{n,n}$. Now note by Cramer's rule that $y_i \stackrel{\circ}{=} \det \tilde{A}_i / \det \tilde{A}_{i+1}$ ($1 \leq i \leq n-1$) is the "x_1" solution to

$$\tilde{A}_{i+1}\vec{x} = \begin{pmatrix} 1 \\ 0 \\ \vdots \\ 0 \end{pmatrix}.$$

Therefore, in time $E_\infty(n)$, the $\{y_i | 1 \leq i \leq n-1\}$ can be found. Finally, det A (= det $\tilde{A}_n$) is computed as $a_{n,n} / \prod_{i=1}^{n-1} y_i$ in $\lceil \log n \rceil$ more steps. □

We can, however, provide for any of these problems no significantly faster algorithm than the following Gauss-Jordan elimination for solving $A\vec{x} = \vec{b}$.

ALGORITHM 6.4.2

Let $A = (a_{ij})_{n \times n}$ $(1 \leq (i,j) \leq n)$ and let $a_{i,n+1}$ be the ith element of the vector $\vec{b}$.
BEGIN
FOR $j := n$ STEP -1 UNTIL 1 DO

$$a_{ik} := a_{ik} \times a_{jj} - a_{ij} \times a_{jk} \ (i \neq j; k \leq j \text{ or } k = n+1)$$

COMMENT: in two steps we eliminate every element of column j except a_{jj};

$$a_{i,n+1} := a_{i,n+1}/a_{ii} \ (1 \leq i \leq n);$$

COMMENT: $a_{i,n+1}$ gives the ith element of the solution;
END.

This algorithm requires $2n + 1$ steps and uses roughly $2n^2$ processors. Indeed, if we were to do the necessary pivoting in order to produce a numerically stable solution, then $O(n \log n)$ steps, including comparisons, would be required. The best lower bound we can produce for this problem is roughly $2 \log n \cdots$, a far cry from $2n + 1$. It is an open question whether or not Gaussian elimination is essentially optimal for the class of problems under discussion. Should this turn out to be the case, it would provide some evidence against general purpose parallel processors of the form we have envisioned. Vector (or pipeline) machines are clearly just as productive (and far less costly from a hardware and software point of view) for this type of computation than is the more general model.

Very frequently, the matrices encountered in practice are of some special form (which may make the computation easier). One particularly common form is the tridiagonal matrix; i.e., $a_{ij} = 0$ unless

$j = i - 1$, i, or $i + 1$. Stone (1973) presents an O(log n) algorithm for solving $A\vec{x} = \vec{b}$, where A is tridiagonal. Algorithm 6.4.3 suggests a technique for finding the determinant of such a matrix. Without loss of generality, we assume A is a matrix of order 2^k (if it is not, then extend to such a dimension by adding enough rows and columns containing only 1's on the diagonal.

ALGORITHM 6.4.3

Write

$$A = \begin{pmatrix} A_1 & \vdots & a' & 0 \\ \cdots & \cdots & \vdots & \cdots \\ & a'' & \vdots & \\ 0 & & \vdots & A_2 \end{pmatrix}$$

where a' denotes $a_{n/2,(n/2)+1}$, a'' denotes $a_{(n/2)+1,n/2}$, and A_1 and A_2 are tridiagonal matrices of order $n/2$. Construct $\overline{A}_1$ from A_1 by replacing elements $(n/2,(n/2)-1))$ and $((n/2)-1,n/2)$ by 0 and element $(n/2,n/2)$ by a'.

Similarly form $\overline{A}_2$ from A_2 by replacing elements $(1,2)$ and $(2,1)$ by 0, and $(1,1)$ by a''. Then det(A) can be found recursively as

$$\det(A) = \det(A_1)\det(A_2) - \det(\overline{A}_1)\det(\overline{A}_2).$$

The time required to solve the $n = 2^0$ case is clearly 0, and the 2^k case uses two steps plus the time for the 2^{k-1} case. It therefore follows that $2\lceil \log n \rceil$ steps are required for this algorithm. From the fan-in argument $\lceil \log 3n \rceil$ steps are needed, and so we are within a factor of 2 of the lower bound. It is clear that separately computing A_1 and $\overline{A}_1$ makes very inefficient use of the processors available. The form of this algorithm is also poor for use on a vector processing machine. We have left as an exercise (Problem 6.8) the reformulation of the algorithm to avoid these difficulties.

Now consider the problem of inverting a (lower) triangular matrix ($a_{ij} = 0$ if $j > i$). Note that while the determinant can trivially be found in $\lceil \log n \rceil$ parallel steps, computing the inverse directly from the adjoint matrix involves finding determinants of Hessenberg matrices. The following technique suggests a method of computing such determinants, but is phrased in terms of (directly) inverting a triangular matrix.

ALGORITHM 6.4.4 (Heller [1973].)

Again assume (without loss of generality) that $n = 2^k$. Write the triangular n by n matrix A as

$$A = \begin{pmatrix} A_{11} & 0 \\ A_{12} & A_{21} \end{pmatrix},$$

where A_{12} is a full $n/2$ by $n/2$ matrix and A_{11} and A_{21} are lower triangular. Then

$$A^{-1} = \begin{pmatrix} A_{11}^{-1} & 0 \\ -A_{21}^{-1} A_{12} A_{11}^{-1} & A_{21}^{-1} \end{pmatrix}$$

suggests finding A^{-1} by inverting A_{11} and A_{21} (in parallel) and then performing the two matrix multiplications to form $A_{21}^{-1} A_{12} A_{11}^{-1}$. If $I'_\infty(n)$ denotes the time required to invert a triangular n by n matrix, then

$$I'_\infty(n) \leq I'_\infty(n/2) + \lceil 2 \log n \rceil$$
$$= O(\log^2 n).$$

PROBLEMS

6.1   Establish a lower bound for n by n matrix multiplication under k-parallelism.

6.2   Outline the scheduling of processors for Dorn's polynomial evaluation scheme so as to essentially achieve the bound of $\lceil 2n/k \rceil + \lceil 2 \log k \rceil$ steps.

6.3   Describe a procedure based on Algorithm 6.2.2 for the evaluation of a polynomial of degree n under k-parallelism in $2n/k + \log k + O((\log k)^{1/2})$ steps, provided $k \geq O(\log n)$.

6.4   Describe and prove the optimality of a polynomial evaluation algorithm under 2-parallelism.

6.5   (Brent, Kuck, and Maruyama [1973]). Describe an algorithm for computing continued fractions of length n in $O(\log n)$ steps.

6.6   (Csanky [1974]). Show that a system of linear equations can be solved in fewer parallel steps than are taken by Gaussian elimination ($2n - 1$).

6.7   (Csanky [1974]). Show that $E_\infty(n) \leq I_\infty(n) + 2$.

6.8   Reformulate Algorithm 6.4.3 so that it could be run on a vector of pipeline machine in time $O(\log n)$.

6.9   (Brent, Kuck, and Maruyama [1973]). Show how to evaluate an arithmetic expression containing n terms but no divisions in $c \log n$ steps ($c < 4$).

6.10   Modify Algorithm 6.4.2 so that only $2n - 1$ steps are required.

CONJECTURES–OPEN PROBLEMS

(1) Can the $(2 \log n)^{1/2}$ term in the bound of Algorithm 6.2.2 be reduced or eliminated?

(2) Can a system of linear equations be solved in $o(n)$—say, $O(\log n)$—parallel steps?

Chapter 7

THE COMPLEXITY OF RATIONAL ITERATIONS

We have discussed the difficulty of computing several types of rational functions. There are, however, many computational instances in which it is necessary to evaluate a square root or a logarithm, find the zeros of a polynomial, or compute some other irrational function. Clearly, this cannot be done under a model permitting only the four basic arithmetic operations. We therefore turn our attention to techniques for computing rational approximations to these functions. The complexity question may first be phrased as, "How much work is it to compute the desired result to n significant figures?" Since our remarks will be confined to iterative techniques, we need a measure of efficiency that is monotonically *increasing* in the precision gained at each step and *decreasing* in the number of operations used at each stage. Limits will be shown on the efficiency with which numbers can be approximated.

7.1 A Measure of Efficiency

Before formalizing a notion of efficiency, let us consider the Newton iteration for square roots. To find $\sqrt{A}$, $A > 0$, let x_0 be any positive number and

$$x_{i+1} = 1/2\,(x_i + A/x_i).$$

Observe that if the sequence $\{x_i\}$ converges, then it converges to $\alpha = \sqrt{A}$. To show that it does converge, note that if $x_0 < \alpha$, then

A Measure of Efficiency 149

$x_1 > \alpha$, and that $x_i > \alpha$ implies $x_{i+1} > \alpha$. The sequence $x_i (i \geq 1)$ is, then, decreasing and bounded below by α. Hence, it has a limit, which must be α. But how quickly does it converge, once it "gets going"? Consider the error at the ith stage, $e_i = |\alpha - x_i|$; then

$$e_{i+1} = |\alpha - 1/2 ((\alpha + e_i) + A/(\alpha + e_i))|$$

$$= |e_i^2 / 2(\alpha + e_i)| = O(e_i^2),$$

(assuming e_i to be small with respect to α). Informally, "the number of significant digits roughly doubles at each step," or the convergence is quadratic.

With this type of computation in mind, let $x_0, x_1, x_2, \ldots$ be a convergent sequence with limit α, and $x_i \neq \alpha$ for "almost all" i. If the sequence $\{x_i\}$ is generated by some iterative method, then $x_{i+1} = \phi(x_i, x_{i-1}, \ldots, x_{i-d})$, where (if the method is computationally realistic) ϕ is a multivariate rational function. Questions of this nature have been extensively studied by Traub (1964), Ostrowski (1966), and others. (See Traub [1972] for a survey.) They analyze various classes of iterative methods for finding a root α of a function f. The measure of complexity is a trade-off between the "order" of convergence of the resulting sequence and the number of evaluations of the function f and its derivatives. The order indicates how many iteration steps will be needed to achieve a desired accuracy, and the number of function evaluations is an indication of the complexity of the iteration step.

Motivated by this work, Paterson (1972) and then Kung (1973a) investigated the efficiency of rational iteration sequences where the complexity of a step is now the number of basic operations (e.g., the number of multiplications). The results of Section 7.2 are initially due to Paterson, with a number of refinements by Kung. We will follow the exposition of Kung (1973a). Throughout this section, all operations are taken with respect to either the field of real or complex numbers.

DEFINITION 7.1.1

Let $e_i = |x_i - \alpha|$. The sequence $\{x_i\} \to \alpha$ has *order $p > 1$* if and only if $\lim_{i\to\infty} \left(e_{i+1}/e_i^{p-\epsilon}\right) = 0$, $\lim_{i\to\infty} \left(e_{i+1}/e_i^{p+\epsilon}\right) \neq 0$ for any $\epsilon > 0$. This yields the more standard notion that

$$p = \sup \left\{ r \mid \lim_{i\to\infty} \frac{e_{i+1}}{e_i^r} = 0 \right\}$$

But also, it immediately follows that the sequence $\{x_{in} \mid n$ a fixed positive integer$\}$ has order p^n.

Let M be the number of basic operations per step that will be "counted." From the preceding remark it is apparent that if we compose a method with itself so as to effectively do two iteration steps as one step, then the order of the "new method" is p^2. However, the work per step will now be $2M$. This motivates the following definition of *efficiency* (Ostrowski [1966]):

$$E \stackrel{\circ}{=} \frac{\log_2 p}{M}$$

Now E (or a strictly increasing function of E such as $p^{1/M}$) will be the same for (the sequence generated by) a composite method $\phi \circ \phi$ as for ϕ. Intuitively, $1/E$ represents the cost of doubling the precision of the approximation.

To summarize, call $\{x_i\}$ a pth order iterative sequence for α generated by ϕ if and only if

(i) $\lim_i x_i = \alpha$.
(ii) $x_i \neq \alpha$ for all but finitely many i.
(iii) $\{x_i\}$ has order $p > 1$ (if $p = 1, E = 0$).
(iv) $x_{i+1} = \phi(x_i, x_{i-1}, \ldots, x_{i-d})$, where $\phi(y_1, \ldots, y_{d+1}) =$

$$\phi_1(y_1, \ldots, y_{d+1}))/(\phi_2(y_1, \ldots, y_{d+1}))$$

and ϕ_1 and ϕ_2 are relatively prime multivariate polynomials.

Throughout this chapter we shall state and prove results only for the case in which ϕ is a function of one variable, although they also hold in the multivariate case.

Bounds on Efficiency

7.2 Bounds on Efficiency

Let M be the number of multiplications or divisions (and $\overline{M}$ the number of nonscalar $*$ operations) needed to compute $\phi(y)$.

THEOREM 7.2.1

$E = (\log p/M) \leq \overline{E} \overset{\circ}{=} (\log p/\overline{M}) \leq 1$ for any iterative sequence generated by a rational method ϕ.

Proof. The basic idea of the proof is fairly straightforward. We will show that a rational iteration function ϕ must have degree $\geq p$ if it is to generate a sequence of order p. By Theorem 2.4.7 or Theorem 5.1.14, the computation of a function ϕ of degree p requires $\overline{M} \geq \lceil \log p \rceil$. Hence,

$$\overline{E} = \frac{\log p}{\overline{M}} \leq \frac{\log p}{\lceil \log p \rceil} \leq 1.$$

To complete the proof we need to verify the initial lemma.

LEMMA 7.2.2

For any α, and any pth order sequence for α generated by ϕ, the degree $\phi \geq p$.

Proof. We will avoid the full detail and sketch only the idea for the special (one-point iteration, i.e., iteration with no memory) case $x_{i+1} = \phi(x_i) = \phi_1(x_i)/\phi_2(x_i)$. We can expand $\phi_1(x) - \alpha\phi_2(x) = \sum_j c_j(x-\alpha)^j$, where the constants c_j do not depend on x (but do depend on α). Suppose that the degree $\phi < p$. Then $\phi_1(x) - \alpha\phi_2(x) = \sum_{j<p} c_j(x-\alpha)^j$. We now show that each $c_j = 0$, and hence $\phi \equiv \alpha$; i.e., ϕ does not generate a nontrivial sequence. By the definition of order, for $\epsilon > 0$:

$$\lim_{i \to \infty} \frac{|x_{i+1} - \alpha|}{|x_i - \alpha|^{p-\epsilon}} = \lim_{i \to \infty} \frac{|\phi(x_i) - \alpha|}{|x_i - \alpha|^{p-\epsilon}} = 0.$$

Then

$$\lim_{i \to \infty} \frac{\left| \sum_{j=0}^{p-1} c_j (x_i - \alpha)^j \right|}{e_i^{p-\epsilon}} = 0.$$

Now, by induction, show $c_j = 0$ for all $j < p$:
- (i) Because $\lim_i e_i \to 0$, $e_i^{p-\epsilon} \to 0$, and therefore $c_0 = 0$.
- (ii) Assume $c_k = 0$ for $k \leq j < p - 1$. Show $c_{j+1} = 0$.

$$\lim_{i \to \infty} \frac{\sum_{k=j+1}^{p-1} c_k (x_i - \alpha)^k / e_i^{j+1}}{e_i^{p-\epsilon} / e_i^{j+1}} = 0$$

As $i \to \infty$, $e_i^{p-\epsilon} / e_i^{j+1} \to 0$ for sufficiently small ϵ, since $j + 1 < p$; thus $c_{j+1} = 0$.

□

The bounds of Theorem 7.2.1 are sharp in that iteration sequences can be found which meet these bounds.

FACT 7.2.3

(1) The Newton iteration for $\alpha = \sqrt{A}$ has efficiency $\overline{E} = 1$, since, as previously noted, it has order 2 and $\overline{M} = 1$.

(2) The iteration $x_{i+1} = x_i^2 + x_i - 1/4$ with limit $-1/2$ has order 2, $M = 1$; and so, $E = 1$.

Part (2) seems uninteresting, especially when compared with (1), until one thinks of that uncounted scalar multiplication in Newton iteration. We now turn to the problem of achieving even sharper bounds on the efficiency of computing algebraic numbers.

DEFINITION 7.2.4

For any algebraic number α, we define the *minimum polynomial* of α, $(m(x))$, as the polynomial of minimal degree with integer coefficients such that $m(\alpha) = 0$. It is well known that if the coefficients of $m(x)$ are relatively prime, then $m(x)$ is unique. The *degree of α* is then the degree of $m(x)$.

Bounds on Efficiency

THEOREM 7.2.5 (Kung [1973b].)

Let α be an algebraic number of degree r; then
(1) $\bar{E}$ can be 1 if and only if $r = 1$ or 2; and
(2) E can be 1 if and only if $r = 1$.

Proof. The sufficiency condition follows from Fact 7.2.3. To show necessity, we must prove several lemmas and an interesting auxiliary theorem (Theorem 7.2.8).

We will use I to denote the integers (either real or Gaussian) and speak of $f(x)$ ($\equiv f_1(x)/f_2(x)$) as a rational function over the integers if it can be written as the ratio of two relatively prime polynomials $f_i \in I[x]$. By convention, if $g(x) \in I(x)$, $g_1(x)$ and $g_2(x) \in I[x]$ will be implicitly defined by $g(x) \stackrel{\circ}{=} g_1(x)/g_2(x)$.

LEMMA 7.2.6

Let $\phi(x) \not\equiv 0$ be a rational function over the integers. If $\phi^{(i)}(\alpha) = 0$ for $i = 0, \ldots, n - 1$ and α is an algebraic number then

$$\phi_1(x) = q(x) \cdot (m(x))^n,$$

where $q(x) \in I[x]$ and $m(x)$ is the minimal polynomial of α.

Proof. We prove the lemma by induction on n. It is known that a polynomial over I with a zero at α is divisible by $m(x)$; hence, the lemma holds for $n = 1$.

Assume the lemma holds for $n \leq k$ and that $\phi^{(i)}(\alpha) = 0$ for $i = 0, \ldots, k$. By the induction hypothesis we may write $\phi_1(x)$ as $\phi_1(x) = p(x) \cdot (m(x))^k$ for a polynomial $p(x) \not\equiv 0$ in $I[x]$. It remains to be shown that $m(x)$ divides $p(x)$. Write

$$\phi(x) = p(x) \cdot r(x) \quad r(x) = m(x)^k/\phi_2(x);$$

then

$$\phi^{(k)}(x) = \sum_{i=0}^{k} \binom{k}{i} p^{(k-i)}(x) \cdot r^{(i)}(x).$$

But as $\phi^{(k)}(\alpha) = 0$ and $r^{(i)}(\alpha) = 0$ for $i = 0, \ldots, k - 1$ follow from the induction hypothesis, we have

$$0 = p(\alpha) \cdot r^{(k)}(\alpha).$$

From the fact that $m(\alpha) = 0$, the determination of $r^{(k)}(\alpha)$ becomes tolerable and indeed

$$r^{(k)}(\alpha) = k!(m'(\alpha))^k/\phi_2(\alpha).$$

Since neither $m'(\alpha)$ nor $\phi_2(\alpha)$ are 0, $r^{(k)}(\alpha) \neq 0$. Therefore $p(\alpha) = 0$, implying $m(x)$ divides $p(x)$. □

LEMMA 7.2.7

If $\phi(x) \in I[x]$ generates a pth order sequence for an algebraic number α, then for $k = 1, \ldots, \lceil p \rceil - 1$, $\phi^{(k)}(\alpha) = 0$.

Proof. From the proof of Lemma 7.2.2 we see that in

$$\phi_1(x) - \alpha\phi_2(x) = \sum_{j=0}^{\deg(\phi)} c_j(x-\alpha)^j \quad c_j = 0 \text{ for } j < \lceil p \rceil.$$

Hence, the kth derivative of the above expression at α is 0. However,

$$\phi^{(k)}(x) = \frac{d^k}{dx^k}(\phi(x) - \alpha) = \frac{d^k}{dx^k}\frac{\phi_1(x) - \alpha\phi_2(x)}{\phi_2(x)}$$

□

The following theorem is not only useful in proving our main result, but also very interesting in its own right.

THEOREM 7.2.8 (Auxiliary of Theorem 7.2.5)

Suppose α is an algebraic number of degree $r \geq 2$ and $\phi(x)$ generates a sequence converging to α. Then

$$\overline{M} \geq \log[r(\lceil p \rceil - 1) + 1] - 1.$$

Proof. Since $\phi(\alpha) = \alpha$, we can choose a neighborhood of α in which $\phi_2(x) \neq 0$; the remainder of this proof will consider only such a

Bounds on Efficiency

neighborhood. From Lemma 7.2.7, $\phi^{(i)}(\alpha) = 0$ for $i = 1, \ldots, \lceil p \rceil - 1$.

If $\psi(x)$ denotes $\phi^{(1)}(x)$, and $\psi_1(x)$ and $\psi_2(x)$ have the usual meanings, then

$$\psi^{(i)}(x) \not\equiv 0, \quad \psi^{(i)}(\alpha) = 0 \quad \text{for } i = 0, \ldots, \lceil p \rceil - 2.$$

Therefore, Lemma 7.2.6 tells us that

$$\deg \psi_1 \geq (\lceil p \rceil - 1) \deg m = (\lceil p \rceil - 1)r.$$

From the definition of ψ and the "derivative of products" rule it follows that

$$2 \deg \phi \geq \deg \psi_1 + 1 \quad \text{or} \quad \deg \phi \geq \tfrac{1}{2} (\lceil p \rceil - 1)r,$$

or

$$\overline{M} \geq \log[r(\lceil p \rceil - 1) + 1] - 1. \qquad \square$$

An alternate statement of this auxiliary theorem is

$$\overline{E} \leq \log p / (\log [r(\lceil p \rceil - 1) + 1] - 1).$$

This result implies that $E < 1$ whenever r and p are both greater than 2; this will be used in the final proof of Theorem 7.2.5. It also tells us that if we are to bound the order of convergence of the algorithms considered (by bounding the number of multiplications per iteration), then $\overline{E}$ as a function of r is at most $O((\log r)^{-1})$. Further comments on evaluating "higher order" algebraic numbers can be found in the problems.

Returning to the development of the main result, consider the following lemma.

LEMMA 7.2.9

Let $\phi(x)$ be a rational function over the integers and α an algebraic number. Then $\phi(\alpha) = \alpha$ and $\phi^{(i)}(\alpha) = 0$ for $i = 1, \ldots, p - 1$ ($p \geq 2$) implies that

$$\phi_1^{(i)}(\alpha) - \alpha\phi_2^{(i)}(\alpha) = 0 \quad \text{for } i = 0, \ldots, p-1$$

Proof. Since $\phi_1(x) = \phi(x) \cdot \phi_2(x)$ we have

$$\phi_1^{(i)}(x) = \Sigma_{0 \leqslant j \leqslant i} \binom{i}{j} \phi^{(j)}(x) \cdot \phi_2^{(i-j)}(x)$$

The hypothesis of the lemma then yields

$$\phi_1^{(i)}(\alpha) = \phi(\alpha) \cdot \phi_2^{(i)}(\alpha) \quad \text{for } i = 0, \ldots, p-1$$
$$= \alpha\phi_2^{(i)}(\alpha).$$

That is,

$$\phi_1^{(i)}(\alpha) - \alpha\, \phi_2^{(i)}(\alpha) = 0 \quad \text{for } i = 0, \ldots, p-1.$$

LEMMA 7.2.10

If $M(\phi) = \log(\deg \phi)$ where $\phi(x) \in I(x)$, then $\deg \phi_2 < \deg \phi_1 = 2^{M(\phi)}$ and the leading coefficient of $\phi_1(x)$ is divisible by that of $\phi_2(x)$.

We omit the proof of this lemma (see Kung [1973b]). It proceeds by induction on M and is similar to the proof of Theorem 2.4.7.

Proof of Theorem 7.2.5

We first complete the proof of part (1) of the theorem; i.e., $\overline{E} = 1$ implies $r = 1$ or 2.

Suppose $\overline{E} = 1$; now if $r > 2$, then by Theorem 7.2.8, $p \leqslant 2$. But it follows directly from $\overline{E} = 1$ and $\overline{M} \geqslant 1$, that $\log p = \overline{M}$ and thus $p = 2, \overline{M} = 1$.

Since $\overline{M} = 1$, $\deg \phi_1 \leqslant 2$ and $\deg \phi_2 \leqslant 1$. Thus, $\phi_1(x) - x\phi_2(x)$ is of degree at most 2. Now $\phi_1(x) - x\phi_2(x) \not\equiv 0$, for if it were, it would imply $\phi(x) \equiv x$ and so $\phi'(x) = 1$; but from Lemma 7.2.7 we know $\phi'(\alpha) = 0$.

On the other hand, $\phi_1(\alpha) - \alpha\phi_2(\alpha) = 0$. This implies α is a zero of a polynomial over I of degree 1 or 2. This completes the proof of this part of the theorem.

To verify that $E = 1$ implies $r = 1$, note that if $E = 1$ then $\overline{E} = 1$, so we need only assume $r = 2$ and argue to a contradiction.

Since $E = 1$, $M = \log p$ and so p is an integer and at least 2. From Lemma 7.2.9,

$$\phi^{(p-1)}(\alpha) - \alpha\phi^{(p-1)}(\alpha) = 0.$$

Note that

(7.2.1) $\qquad \phi_1^{(p-1)}(x) - x\phi_2^{(p-1)}(x)$

is a polynomial of degree at most 1. Suppose it is identically 0. Let a_1 denote the lead coefficient of $\phi_1(x)$; and a_2, the lead coefficient of $\phi_2(x)$. Then $pa_1 = a_2$ ($p \geq 2$), but this cannot be because Lemma 7.2.10 says a_1 is divisible by a_2. Therefore (7.2.1) is in fact of degree 1 and α is a root. This implies that α is an algebraic number of degree 1, contradicting the assumption $r = 2$. Therefore, $r = 1$.

□

This concludes not only the proof of Theorem 7.2.5, but also the section, chapter, and the monograph itself. We have spared the reader what would have been a content-free summary.

PROBLEMS

7.1 Extend the proof of Lemma 7.2.2 to the multivariate case (i.e., ϕ is a multivariate rational function).

7.2 Show that the square root function may be computed with efficiency $E = 1 - \epsilon$ (for any $\epsilon > 0$).

7.3 Give an iterative scheme for computing cube roots. Prove that it converges to the proper result. What is its efficiency?

7.4 (Paterson [1972].) Describe an iterative scheme for computing $\sqrt[n]{A}$ (n large, A in $\mathbf{Q}$) with efficiency $\overline{E} \simeq 2 \log \log n/\log n$.

7.5 (Strassen [1973c].) Let $\alpha \in \mathbf{R}$ and $\epsilon > 0$. Given $\{0,1\}$ and the operations $+, -, \times, \div$, we want to compute some $\beta \in \mathbf{R}$ satisfying $|\beta - \alpha| < \epsilon$. Let $L_\epsilon(\alpha)$ be the minimum number of operations to compute a suitable β. Let $C(\epsilon) = \log(1/\epsilon)/\log\log(1/\epsilon)$; i.e., for $\epsilon = 2^{-n}$, $C(\epsilon) = n/\log n$.

(a) By a counting argument, show that for "most" $\alpha \in \mathbf{R}$, $\lim_{\epsilon \to 0} [L_\epsilon(\alpha)/C(\epsilon)] = 1$.
(b) Show that for all $\alpha \in \mathbf{R}$, $\lim_{\epsilon \to 0} \sup[L_\epsilon(\alpha)/C(\epsilon)] \leq 1$.
(c) Show that for all algebraic $\alpha \in \mathbf{R}$, $L_\epsilon(\alpha) = O(\log \log 1/\epsilon)$.

CONJECTURES–OPEN PROBLEMS

1. When iterating to find a root of an nth degree polynomial, must the efficiency E (or $\bar{E}$) $\to 0$ as $n \to \infty$?

2. (See Problem 7.5.) Find a specific transcendental $\alpha \in \mathbf{R}$ (e.g., $\alpha = e = 2.718\ldots$, or $\alpha = \pi$) for which $\lim_{\epsilon \to 0} \sup[L_\epsilon(\alpha)/C(\alpha)] > 0$.

GLOSSARY OF MATHEMATICAL NOTATION

Notation	Brief explanation
$\log n$	$\log_2 n$, all logarithms are taken to base 2 unless otherwise stated.
$O(f(n))$	$b(n) = O(f(n))$ denotes the fact that $b(n)$ grows "no more rapidly" than $f(n)$; i.e., $\lim_{n \to \infty} \sup b(n)/f(n) < c$ for some constant c. In particular, we may use $O(1)$ to denote a function bounded by a constant.
$\lceil a \rceil$	the least integer greater than or equal to a; we also use $\lfloor a \rfloor$ to dentoe the greatest integer less than or equal to a.
$F[x_1, \ldots, x_n]$	The ring extension to the field F by the elements $x_1, \ldots, x_n$; i.e., the ring of polynomials in $x_1, \ldots, x_n$ over F.
$F(x_1, \ldots, x_n)$	The field extension to the field F by the elements $x_1, \ldots, x_n$; i.e., the field of rational functions in $x_1, \ldots x_n$ over F.
R	the field of real numbers
C	the field of complex numbers
Q	the field of rational numbers

Z	the ring of integers
$\mathbf{Z}_p$	the ring of integers modulo p.
$f(x_1, \ldots x_n)\|_{x_n \leftarrow y}$	the function, $f'(x_1, \ldots, x_{n-1})$ substituting y for x_n in f.
$P\|_{x \leftarrow y}$	the program P' formed by substituting y for x in program P.
$\vec{a}$	the n-tuple $\langle a_1, \ldots, a_n \rangle$ used as an abbreviation for function arguments; e.g., $f(\vec{a})$ for $f(a_1, \ldots, a_n)$; $\vec{a}$ is also used to denote a vector.
N	the nonnegative integers.
*	the operation $\times$ (multiplication) or $\div$ (division); also complex conjugate when appearing as a superscript.
$\stackrel{\circ}{=}$	"is defined as."
$\mathbf{Q}\langle a_1, \ldots, a_k \rangle$	the set of *generalized monomials* over $\{a_1, \ldots, a_k\}$; i.e., the set of terms of the form $\prod_{i=1}^{k} a_i^{c_i}$ with $c_i \in \mathbf{Q}$.
$\vec{\phi}_{1 \times n}$	the column vector $$(\phi_1, \ldots, \phi_n)^T = \begin{pmatrix} \phi_1 \\ \vdots \\ \phi_n \end{pmatrix}$$
$\Phi_{n \times m}$	the n by m matrix Φ.
$L_F(\vec{a}, G)$	the set of linear affine functions over F; i.e., $\left\{ \sum_{i=1}^{n} c_i a_i + g \mid c_i \in F, g \in G \right\}$.

Glossary

$F[[x_1,\ldots,x_n]]$ — the power series ring generated by $x_1,\ldots,x_n$ over the field F; i.e., elements of the form

$$\sum_{\langle i_1,\ldots,i_n\rangle \,\in\, \mathbf{N}^n} c_{i_1\cdots i_n} x_1^{i_1}\cdots x_n^{i_n}$$

with $c_{i_1\cdots i_n} \in F$.

$\langle\!\langle x,y\rangle\!\rangle$ — the vector inner product $\Sigma x_i y_i$.

$\mathrm{o}(f(n))$ — $b(n) = \mathrm{o}(f(n))$ denotes the fact that $b(n)$ grows more slowly than $f(n)$; i.e., $\lim_{n\to\infty} \sup b(n)/f(n) = 0$.

a^* — complex conjugate of a.

A^* — conjugate transpose of the matrix A.

BIBLIOGRAPHY AND REFERENCES

Aanderaa, S. O., see Cook and Aanderaa (1969).
Aho, A. V., Hopcroft, J. E., and Ullman, J. D. (1974) *The Design and Analysis of Computer Algorithms*, Addison Wesley, Reading, Mass.
———, Steiglitz, K., and Ullman, J. D. (1975) "Evaluating Polynomials at Fixed Sets of Points", to appear *SIAM Journal on Computing*.
105
Allen, J., see Moenck and Allen (1974).
Belaga, E. G., (1958) "Some Problems in the Computation of Polynomials", *Dokl. Akad. Nauk. SSSR*, Vol. 123: pp. 775-777.
61
Birkhoff, G., and MacLane, S. (1953) *A Survey of Modern Algebra*, MacMillan, New York.
120
Borodin, A., see also Moenck and Borodin (1972), Munro and Borodin (1972).
———, (1971) "Horner's Rule is Uniquely Optimal", in *Theory of Machines and Computation*, Academic Press, New York.
———, (1973a) "Computational Complexity—Theory and Practice", in A. Aho (Ed.), *Currents in the Theory of Computing*, Prentice Hall, Englewood Cliffs, N. J., pp. 35-89.
1
———, (1973b) "On the Number of Arithmetics to Compute Certain Functions —Circa May 1973", in J. F. Traub (Ed.), *Complexity of Sequential and Parallel Numerical Algorithms*, Academic Press, New York, pp. 149-180.
———, and Cook, S. (1974) "On the Number of Additions to Compute Specific Polynomials", *Proc. 6th Annual ACM Symp. on Theory of Computing*, pp. 342-347.
68, 117, 121
———, and Moenck, R., (1974) "Fast Modular Transforms", *Journal of Computer and Systems Sciences*, vol. 8, no. 3, pp. 366-386.
99
———, and Munro, I. (1971) "Evaluation of Polynomials at Many Points", *Information Processing Letters*, vol. 1, no. 2, pp. 66-68.
Brauer, A. (1939) "On Addition Chains", *Bull. AMS*, vol. 45, pp. 736-739.
126

Brent, R. P. (1970) "Algorithms for Matrix Multiplication", Stanford University Report CS157.
133

———, (1973) "The Parallel Evaluation of Arithmetic Expressions in Logarithmic Time", in J. F. Traub (Ed.), *Complexity of Sequential and Parallel Numerical Algorithms,* Academic Press, New York, pp. 83-102.
142

———, (1974) "The Parallel Evaluation of General Arithmetic Expressions", *JACM,* vol. 21, no. 2, April, pp. 201-206.
137-141

———, Kuck, D. J., Maruyama, K. M. (1973) "The Parallel Evaluation of Arithmetic Expressions without Division", *IEEE Transactions on Computers* C-22, May, pp. 532-534.
142, 147

Brockett, R. and Dobkin, D. (1973) "On the Optimal Evaluation of a Set of Bilinear Forms", *Proc. 5th Annual ACM Symp. on Theory of Computing,* pp. 88-95.
35, 39, 43

Bunch, J. and Hopcroft, J. (1974) "Triangular Factorization and Inversion by Fast Matrix Multiplication", *Math. of Computation,* vol. 28, no. 125, pp. 231-236.
49-50

Chevonenkis, O., see Lyusternik, Chervonenkis and Yanpol' skii (1965).

Collins, G. E. (1971) "The SAC-1 System: An Introduction and Survey", *Proc. of 2nd Symp. on Symbolic and Algebraic Computation,* ACM, New York, pp. 144-152.
78, 88

Cook, S. A., see also Borodin and Cook (1974).

———, (1971) "The Complexity of Theorem-Proving Procedures", *Proc. 3rd Annual ACM Symp. on Theory of Computing,* pp. 151-158.
2

Cook, S. A. and Aanderaa, S. O. (1969) "On the Minimum Computation of Functions", *Trans. American Math. Society,* vol. 142, pp. 291-314.
104

Cooley, J. W., Lewis, P. A. and Welch, P. D. (1967) "History of the Fast Fourier Transform", *Proc. IEEE,* vol. 55, pp. 1675-1677.
77, 82

———, and Tukey, J. (1965) "An Algorithm for the Machine Calculation of Complex Fourier Series", *Math. Comp.,* vol. 19, pp. 297-301.
77

Csanky, L. (1974) "On the Parallel Complexity of Some Computational Problems", Ph.D. Thesis, Dept. EECS, University of Calif. at Berkeley.
143, 147

Bibliography and References 165

Dobkin, D., see also Brockett and Dobkin (1973).

———, (1973) "On the Arithmetic Complexity of a Class of Arithmetic Computations", Yale University, Dept. of Computer Science Research Report 23.
39

Dorn, W. S. (1962) "Generalizations of Horner's Rule for Polynomial Evaluation", *IBM Journal of Research and Development*, no. 4, pp. 239-245.
132

Erdös, P. (1960) "Remarks on Number Theory III—On Addition Chains", *Acta Arithmetica*, vol. 6, pp. 77-81.
26, 126

Estrin, G. (1960) "Organization of Computer Systems—The Fixed Plus Variable Structure Computer", *Proc. Western Joint Conference*, no. 5, pp. 33-40.
132

Eve, J. (1964) "The Evaluation of Polynomials", *Numerische Mathematik*, vol. 6, pp. 17-21.
56

Fiduccia, C. M. (1971) "Fast Matrix Multiplication", *Proc. 3rd Annual ACM Symposium on Theory of Computation*, pp. 45-49.
23, 25

———, (1972a) "On Obtaining Upper Bounds on the Complexity of Matrix Multiplication", in R. E. Miller and J. W. Thatcher (Eds.), *Complexity of Computer Computations*, Plenum Press, New York, pp. 31-40.
39, 43, 44, 52

———, (1972b) "Polynomial Evaluation via the Division Algorithm: The Fast Fourier Transform Revisited", *Proc. 4th Annual ACM Symp. on Theory of Computing*, pp. 88-93.
39, 43, 99, 100

Fischer, P. C. (1974) "Further Schemes for Combining Matrix Algorithms", in J. Loeck (Ed.), *Lecture Notes in Computer Science*, vol. 14, Springer Verlag, pp. 428-436.
45

Ford, L., and Johnson, S. (1959) "A Tournament Problem", *Amer. Math. Monthly*, vol. 66, pp. 391-395.
80

Fulton, W. (1969) *Algebraic Curves*, Benjamin Inc., New York.
109

Gastinel, N. (1971) "Sur les Calcul des products de matrices", *Numerische Mathematik*, vol. 17, pp. 222-229.
39, 43

Gentleman, M. and Sande, G. (1966) "Fast Fourier Transforms—for Fun and Profit", *Proc. Fall Joint Computer Conference*, pp. 563-578.
77

Godement, R. (1968) *Algebra,* Houghton Mifflin, Boston.
60

Hall, A. D. (1971) "The ALTRAN System for Rational Function Manipulation —A Survey", *Proc. of 2nd Symp. on Symbolic and Algebraic Manipulation.* ACM, New York, pp. 153-157.
88

Hardy, G. H. and Wright, E. M. (1960) *An Introduction to the Theory of Numbers* (4th ed.), Oxford University Press, London.
86

Heller, D. (1973) "A Determinant Theorem with Applications to Parallel Algorithms". Dept. of Computer Science Tech. Report, Carnegie-Mellon University.
146

Hopcroft, J. E., see also Aho, Hopcroft and Ullman (1974), Bunch and Hopcroft (1974).

___, and Kerr, L. (1971) "On Minimizing the Number of Multiplications Necessary for Matrix Multiplication", *SIAM Journal on Appl. Math.,* vol. 20, no. 1, pp. 30-36.
24, 39, 45

___, and Musinski, J. (1973) "Duality Applied to the Complexity of Matrix Multiplication and Other Bilinear Forms", *SIAM Journal on Computing,* vol. 2, no. 3, pp. 159-173.
39

Horowitz, E. (1971) "Modular Arithmetic and Finite Field Theory: A Tutorial", *Proc. of 2nd Symp. on Symbolic and Algebraic Manipulation,* ACM, New York, pp. 188-194.
78

___, (1972) "A Fast Method for Interpolation Using Preconditioning", *Information Processing Letters,* vol. 1, no. 4, pp. 157-163.
99, 101, 103

Hyafil, L. and Kung, H. T. (1974) "The Complexity of Parallel Evaluation of Linear Recurrences", Carnegie-Mellon University Computer Science Tech. Report.
142

Ishibashi, Y., see Takahashi and Ishibashi (1961).

Johnson, S., see Ford and Johnson (1959).

Karp, R. M. (1971) "Lecture Notes on Specific Complexity Theory", unpublished.
88

___, (1972) "Reducibility among Combinatorial Problems" in R. E. Miller and J. W. Thatcher (Eds.), *Complexity of Computer Computations,* Plenum Press, New York, pp. 85-104.
2

Bibliography and References

Kedem, Z. (1974) "Combining Dimensionality and Rate of Growth Arguments for Establishing Lower Bounds on the Number of Multiplications", *Proc. 6th Annual ACM Symp. on Theory of Computing,* pp. 334-341.
28

____,and Kirkpatrick, D. (1974) "Addition Requirements for Rational Expressions", unpublished manuscript.
12, 20

Kerr, L., see Hopcroft and Kerr (1971).

Kirkpatrick, D., see also Kedem and Kirkpatrick (1974).

____, (1972a) "On the Additions Necessary to Compute Certain Functions", *Proc. 4th Annual ACM Symp. on Theory of Computing,* pp. 94-101.
52

____, (1972b) "Active Multiplications for Independent Forms", unpublished manuscript.
24

Knuth, D. E. (1969) *The Art of Computer Programming: Seminumerical Algorithms,* vol. II, Addison Wesley, Reading, Mass.
26, 55, 57, 75, 77, 78, 97, 103, 132

____, (1973a) *The Art of Computer Programming: Sorting and Searching,* vol. III, Addison Wesley, Reading, Mass.
2

____,(1973b) *The Art of Computer Programming: Seminumerical Algorithms,* vol. II revised. Addison Wesley, Reading, Mass.
2, 94

Kogge, P. M. (1972) "Parallel Algorithms for the Efficient Solution of Recurrence Problems" Tech. Report 314, Digital Systems Laboratory, Stanford University.
129

____,and Stone, H. S. (1972) "An Algorithm for the Parallel Evaluation of a Class of Recurrence Relations", Tech. Report 17, Digital Systems Laboratory, Stanford University.
129

Kuck, D. J., see Brent, Kuck and Maruyama (1973).

Kung, H. T., see also Hyafil and Kung (1974).

____, (1973a) "A Bound on the Multiplication Efficiency of Iteration", *Journal of Computer and System Sciences,* vol. 7, no. 4, pp. 334-342.
149

____, (1973b) "The Computational Complexity of Algebraic Numbers", *Proc. of 5th Annual ACM Symp. on Theory of Computing,* pp. 152-159.
26, 153-156

____, (1973c) "On Computing Reciprocals of Power Series", Carnegie-Mellon Computer Sci. Tech. Report.
95

_____, (1974) "Some Complexity Bounds for Parallel Computation", *Proc. of 6th Annual ACM Symp. on Theory of Computing*, pp. 323-333.
126-129

_____,and Traub, J. F. (1974) "Computational Complexity of One Point and Multipoint Iteration", in R. Karp (Ed.), *Proc. of SIAM-AMS Symp. on the Complexity of Real Computation*.
129

Lewis, P. A., see Cooley, Lewis and Welch (1967).

Lipson, J. D. (1968) "Symbolic Methods for the Computer Solution of Linear Equations with Applications to Flowgraphs", in R. G. Tobey (Ed.), *Proc. of 1968 Summer Institute on Symbolic Mathematical Computation*, IBM Boston Programming Center.
87

_____, (1971) "Chinese Remainder and Interpolation Algorithms", *Proc. of 2nd Symp. on Symbolic and Algebraic Manipulation*, ACM, New York, pp. 372-391.
79, 102

_____, (1974) "The Role of the FFT in Algebraic Manipulation", unpublished manuscript.
86, 87, 88

Lyusternik, L., Chervonenkis, O. and Yanpol' skii, A. (1965) *Handbook for Computing Elementary Functions*, Pergamon, Oxford.
55

MacLane, S., see Birkhoff and MacLane (1953).

Maruyama, K. M., see also Brent, Kuck and Maruyama (1973).

_____, (1973) "On the Parallel Evaluation of Polynomials", *IEEE Trans. on Computers*, C-22, pp. 2-5.
133

Moenck, R., see also Borodin and Moenck (1974).

_____, (1973) "Studies in Fast Algebraic Algorithms", University of Toronto Computer Science Tech. Report 57.
94, 116

_____,and Allen, J. (1974) "Fast Division Algorithms in Euclidean Domains", University of Waterloo Computer Science Tech. Report.
94.

_____,and Borodin, A. (1972) "Fast Modular Transformations via Division", *Proc. 13th Annual IEEE Symp. on Switching and Automata Theory*, pp. 90-96.
95, 99

Morgenstern, J. (1973a) "Note on a Lower Bound of the Linear Complexity of the Fast Fourier Transform", *JACM*, vol. 20, no. 2, pp. 305-306.
117

Bibliography and References 169

_____, (1973b) "Algorithmes Lineaires Tangents et Complexite", *C. R. Acad. Sc. Paris*, t. 277, Serie A, pp. 367-369.
53, 117, 123

Motzkin, T. S. (1955) "Evaluation of Polynomials and Evaluation of Rational Functions", *Bull. Amer. Math. Soc.*, vol. 61, p. 163.
2, 54, 56, 60, 75

Munro, I., see also Borodin and Munro (1971).

_____, (1971a) "Some Results Concerning Efficient and Optimal Algorithms", *Proc. 3rd Annual ACM Symp. on Theory of Computing*, pp. 40-44.
25

_____, (1971b) "Efficient Determinantion of the Transitive Closure of a Directed Graph", *Information Processing Letters*, vol. 1, no. 2, pp. 56-58.
45

_____, (1973) "Problems Related to Matrix Multiplication", in R. Rustin (Ed.), *Courant Institute Symposium on Computational Complexity*, Algorithmics Press, New York, pp. 137-152.
51

_____, and Borodin, A. (1972) "Efficient Evaluation of Polynomial Forms", *Journal of Computer and System Sciences*, vol. 6, no. 6, pp. 625-638.

_____, and Paterson, M. (1973) "Optimal Algorithms for Parallel Polynomial Evaluation", *Journal of Computer and System Sciences*, vol. 7, no. 2, pp. 189-198.
129, 133-136

Musinski, J., see Hopcroft and Musinski (1973).

Muraoka, Y. (1971) "Parallelism Exposure and Exploitation in Programs", Report No. 424, Department of Computer Science, University of Illinois at Urbana-Champaign.
133

Ostrowski, A. M. (1954) "On Two Problems in Abstract Algebra Connected with Horner's Rule", *Studies Presented to R. von Mises*, Academic Press, New York, pp. 40-48.
2, 8

_____, (1966) *Solution of Equations and Systems of Equations*, Academic Press, New York.
149-150

Pan, V. Y. (1966) "Methods of Computing Values of Polynomials", *Russian Mathematical Surveys*, vol. 21, no. 1, pp. 105-136.
8, 12, 55

Paterson, M., see also Munro and Paterson (1973).

_____, (1972) "Efficient Iterations for Algebraic Numbers", in R. E. Miller and J. W. Thatcher (Eds.), *Complexity of Computer Computations*, Plenum Press, New York, pp. 41-52.
149, 157

_____,and Stockmeyer, L. (1973) "On the Number of Nonscalar Multiplications Necessary to Evaluate Polynomials", *SIAM Journal of Computing*, vol. 2, no. 1, pp. 60–66.
57, 65-67, 76, 104

Pollard, J. (1971) "The Fast Fourier Transform in a Finite Field", *Math. Comp.*, vol. 25, no. 114, pp. 365-374.
88

Probert, R. (1973) "On the Complexity of Matrix Multiplication", University of Waterloo Computer Science Tech. Report CS-73-27.
39, 46

Rabin, M. and Winograd, S. (1971) "Fast Evaluation of Polynomials by Rational Preparation", IBM Tech. Report RC 3645.
57, 59

Revah, L. (1973) "On the Number of Multiplications/Divisions for Evaluating a Polynomial with Auxiliary Functions", M. Sc. thesis, Technion, Israel.
55, 75

Sande, G., see Gentleman and Sande (1966).

Schneider, J. (1957) *Einfuhrung in die Transzendenten Zahlen*, Springer Verlag, Berlin.
74

Schönhage, A. (1971) "Schnelle Berechnung von Kettenbruchentwicklugen", *Acta Informatica*, vol. 1, no. 1, pp. 139-144.
116

_____,(1973a) "Fast Schmidt Orthogonalization and Unitary Transformations of Large Matrices", in J. F. Traub (Ed.), *Complexity of Sequential and Parallel Numerical Algorithms*, Academic Press, New York, pp. 283-291.
50

_____,(1973b) "Eine untere Schranke für die Lange von Additionsketten". To appear in *Theoretical Computer Science*.
26

_____,and Strassen, V. (1971) "Schnelle Multiplikation Grosser Zahlen", *Computing*, vol. 7, pp. 281-292.
77, 88, 94

Shafarevitch, I. (1969) "Foundations of Algebraic Geometry", *Russian Math. Surveys*, vol. 24, no. 6, pp. 1-178.
109, 111

Shaw, M. and Traub, J. F. (1974) "On the Number of Multiplications for the Evaluation of a Polynomial and some of Its Derivatives", *JACM*, vol. 21, no. 1, pp. 161-167.
32, 106

Sieveking, M. (1972) "An Algorithm for Division of Power Series", *Computing*, vol. 10, pp. 153-156.
52, 95

Steiglitz, K., see Aho, Steiglitz and Ullman (1975).
Stockmeyer, L., see Paterson and Stockmeyer (1973).
Stone, H. S., see also Kogge and Stone (1972).
____, (1973) "An Efficient Parallel Algorithm for the Solution of a Tridiagonal System of Equations", *JACM*, vol. 20, no. 1, pp. 27-38.
145
Strassen, V., see also Schönhage and Strassen (1971).
____, (1969) "Gaussian Elimination is Not Optimal", *Numerische Mathematik*, vol. 13, pp. 354-356.
2, 45-46, 49, 78
____, (1972) "Evaluation of Rational Functions", in R. E. Miller and J. W. Thatcher (Eds.), *Complexity of Computer Computations*, Plenum Press, New York, pp. 1-10.
18, 39, 52, 95, 113
____,(1973a) "Die Berechnungskomplexetät von elementarysymmetrischen Funktionen und von Interpolationskoeffizienten", *Numerische Mathematik*, vol. 20, no. 3, pp. 238-251.
23, 108, 123
____, (1973b) "Vermeidung von Divisionen", *J. Reine Angew. Math.*, vol. 264, pp. 184-202.
36, 52
____, (1973c) "Berechnungen in partiellen Algebren endlichen Typs", *Computing*, vol. 11, no. 3, pp. 181-196.
157
____, (1974) "Polynomials with Rational Coefficients which are Hard to Compute", *SIAM Journal on Computing*, vol. 3, no. 2, pp. 128-149.
63, 71
____, (1975) "The Computational Complexity of Continued Fractions", unpublished manuscript.
116
Takahashi, H. and Ishibashi, Y. (1961) "A New Method for Exact Calculation by a Digital Computer", in *Information Processing in Japan*, vol. 1, pp. 28-42.
78
Todd, J. (1955) "Motivation for Working in Numerical Analysis", *Comm. on Pure and Applied Mathematics*, vol. 8, pp. 97-116.
54
Traub, J. F., see also Kung and Traub (1974), Shaw and Traub (1974).
____, (1964) *Iterative Methods for the Solution of Equations*, Prentice-Hall, Englewood Cliffs, N. J.
149
____, (1972) "Computational Complexity of Iterative Processes", *SIAM Journal on Computing*, vol. 1, pp. 167-179.
149

Tukey, J., see Cooley and Tukey (1965).
Ullman, J. D., see Aho, Hopcroft and Ullman (1974) and Aho, Steiglitz and Ullman (1975).
van der Waerden, B. L. (1964) *Modern Algebra*, vol. 1, Frederick Ungar, New York.
60, 94
Valiant, L. (1974) "General Context-Free Recognition in Less Than Cubic Time", Carnegie-Mellon University Computer Science Tech. Report.
45
Vari, T. M. (1972) "On the Number of Multiplications Required to Compute Quadratic Functions", Cornell University Computer Science Tech. Report 72-120.
20
Welch, P. D., see Cooley, Tukey and Welch (1967).
Winograd, S., see also Rabin and Winograd (1971).
____, (1967) "On the Number of Multiplications Required to Compute Certain Functions", *Proc. National Acad. Sci.*, vol. 58, pp, 1840-1842.
2
____, (1969) "On the Algebraic Complexity of Inner Product", IBM Research Report RC 2729.
76
____, (1970a) "On the Number of Multiplications Necessary to Compute Certain Functions", *Comm. on Pure and Applied Mathematics*, vol. 23, pp. 165-179.
6, 15, 35, 38, 52, 59, 131
____, (1970b) "The Algebraic Complexity of Functions", *Actes Congres. Intern. Math.*, vol. 3, pp. 283-288.
____, (1971) "On Multiplication of 2 X 2 Matrices", *Linear Algebra and Its Applications*, vol. 4, pp. 381-388.
25, 46
____, (1972) "Parallel Iteration Methods", in R. E. Miller and J. W. Thatcher (Eds.), *Complexity of Computer Computations*, Plenum Press, New York, pp. 53-60.
129
____, (1973) "Some Remarks on Fast Multiplication of Polynomials", in J. F. Traub (Ed.), *Complexity of Sequential and Parallel Numerical Algorithms*, Academic Press, New York, pp. 181-196.
35
____, (1974) "On Parallel Evaluation of Certain Arithmetic Expressions", IBM Research Report RC 4803.
142
Wright, E. M., see Hardy and Wright (1960).
Yanpol' skii, A., see Lyusternik, Chervonenkis and Yanpol' skii (1965).

INDEX

active operation (multiplication, division), 10-12, 17-34, 51-52
additions (lower bounds), 12-15, 20-23, 35-36, 52-53, 61-64, 67-76, 116-124
affine dimension (adim), 13
a-indep, 21
algebraic dependence, 54-76
algebraic subset of P^N, 110
asymptotic behavior, 5
bilinear form, 34-45, 51-52
β-normal sequence, 32
Chinese remainder theorem, 78
Chinese remaindering process, 101-104
closed subset of P^N, 110
commutativity (non-commutativity), 2-5, 38
complex number multiplication, 25, 40-41
computational complexity, 1
degree of
    algebraic number, 152
    DEG of rational function, 129
    rational function, 127
    set of polynomials, 108
derivative (of polynomial), 32-33, 52, 106
determinant, 51, 142-147
dimension (in projective space), 111
discrete Fourier transform, see fast Fourier transform
divide and conquer, 79-80
division of integers, 95-98
    of polynomials, 95-97

efficiency (of an iterative process), 150
extrapolative recursion, 94-98
fast Fourier transform (FFT), 77-106, 116-120
Gaussian elimination, 49, 87, 142-147
generalized monomial, 13
graph (of set of rational functions), 112
greatest common denominator (GCD), 116
height (of a polynomial), 72
homogenized polynomial, 112
Horner's rule, 8
hyperplane (of P^N), 110
hypersurface (of P^N), 110
inactive operation, 10
interior (of a subspace), 22
interpolation (of polynomials), 81, 101-104, 115
inverse Fourier transform, 84-85
inverse of a matrix, 49-51, 142-147
irreducible set (component), 110
iterative techniques, 7, 148-158
k-parallelism, 7, 125
Lagrangian interpolation, 101-102
LDU factorization of a matrix, 49
linear equations, 142-147
linearly independent (over F mod H), 16
linear program, 117
matrix inversion, 49-51, 142-147
matrix multiplication, 2-5, 23-25, 34-53, 59, 130-131, 142-147

matrix-vector product, 15–25, 76, 130–131
minimum polynomial, 152
model of computation, 5–7
modular arithmetic, 78–80, 87–94
modular representation, 78–80
monomials (generalized), 13
multiplication (see also active operations and nonscalar multiplications)
    of complex numbers, 25, 40–41
    of integers, 88–94
    of matrices, 2–5, 23–25, 34–53, 59, 130–131, 142–147
    of polynomials, 42, 81, 86
Newton iteration, 80, 97, 129, 148–149, 152
nonscalar multiplications, 26–53, 64–71, 76, 81, 115–116
NP complete problems, 2
order of a sequence, 150
parallel computation
    k-parallelism, 7, 125–126, 135–137, 141–142, 146
    linear recurrences, 128–129
    lower bounds, 127–129, 136–137, 142, 146–147
    matrix problems, 130–131, 142–147
    model, 7, 125–126
    polynomial evaluation, 132–137, 146–147
    powers of numbers, 126–127
    rational functions, 137–142, 147
    unbounded parallelism, 7, 126
parse tree (of arithmetic expression), 138
point (in projective P^N space), 109

polynomial evaluation
    from coefficients, 8–15, 28–34, 37–38, 51–53
    with preconditioning, 54–76
    with rational coefficients, 64–71
    of specific polynomials, 71–75, 76
    at many points, 47–49, 83–84, 98–101, 115
    under parallelism, 132–137, 146–147
polynomial multiplication, 42, 81, 86
power series ring, 36
precision of a number (polynomial), 94
preconditioning, 54–59, 75–76
primitive root of unity, 82
projective space, 109
rank (of a matrix or tensor), 42
rational iteration, 7, 148–158
reciprocal of integer or polynomial, 95–98
scalar multiplication, 10, 116–124
square root, 148–149
Strassen's algorithm, 3
substitution argument, 9
symmetric functions, 100, 115
synthetic division, 8
tensor, 39, 42–43
Toeplitz matrix, 42
tradeoff, 3, 75
transcendence basis (degree), 59
triangular matrix, 146
tridiagonal matrix, 145
Turing machine, 1
unbounded parallelism, 7, 126
unit of a power series ring, 36

NOTES

NOTES

NOTES

NOTES

NOTES

NOTES